THE FUTURE OF
MAN

THE WORLD BEFORE THERE WAS LIFE

Billions of years ago, volcanic activity was widespread, but the surface of the earth had cooled sufficiently to permit oceans to form. So far as is known, life had not yet begun. (From a painting by Charles R. Knight in the Ernest Robert Graham Hall of the Chicago Natural History Museum.)

THE FUTURE OF
MAN

By

ROBERT KLARK GRAHAM

Introduction by Professor Sir Cyril Burt

THE CHRISTOPHER PUBLISHING HOUSE
NORTH QUINCY, MASSACHUSETTS

PRINTED IN
THE UNITED STATES OF AMERICA

To my children, in whom I see myself reborn
and, thanks to their mother,
perceptibly improved.

— R.G.

In the preparation of this book, the writer has had the generous counsel of:

General Wayne Bailey
Professor Sir Cyril Burt, D.Litt., Ll.D., D.Sc.
Professor Raymond B. Cattell, Ph.D., D.Sc.
Frank L. Girard, M.S.
Marta Everton, M.D.
Professor Samuel J. Holmes, Ph.D.
Professor John Horn, Ph.D.
Professor Hermann J. Muller, Ph.D., D.Sc.
Maurice W. Nugent, M.D.
William D. Sellers, B.Sc., J.D.

The willingness of these exceptionally able people to assist in a constructive project does not imply their agreement with every part of the book.

CONTENTS

ILLUSTRATIONS AND CHARTS

SUMMARY

The purpose of this book is to describe, as accurately as the available facts permit, the condition from which man has come and why he has come thus far, where he now stands and what major influences are at work upon him. Its further purpose is to set forth how he may alter certain of these major influences so as to realize far more the tremendous potential which reposes within him.

The book traces the gradual development of man into the most intelligent and most dominant creature ever known on Earth. Then it shows how, as man gained control over the hazards of nature, he ceased to be improved further in general intelligence. Thirteen thousand years ago there were less than a million people in the world, but a large proportion of these were rigorously selected and highly intelligent. Now there are billions, but the great majority have mediocre minds or worse. Over half of the dullest people surviving today would have been eliminated under natural selection, while over half of the children who would prove the most effective and valuable citizens are withheld from us through the practice of birth control by the intelligent.

Although the contributions of the intelligent (control of disease, efficient food production and a multitude of inventions) permit great masses of the less intelligent to survive, when these masses gain sufficient preponderance they no longer remain part of a united people. Instead, they choose class war. They are led to turn on the "haves" of their own people and "liquidate" them by the millions. The killing includes not only those who have property, but those who have perceptive minds. Intelligence has come to have extinction potential instead of survival value.

Steps to deal with this vast human derangement are described. These include techniques to enable the unintelligent successfully to limit their births to the number desired. They

13

also include ways to encourage a substantial increase of good minds to cope with man's intensifying problems.

A progressively transforming principle is derived which could raise the average intelligence of all mankind and especially that of any group or nation which elected to take advantage of it. The heart of this simple but elevating principle is: *The more intelligent you are the more children you should have*. Think upon this. Its wide realization could bring stability and benefit to all of mankind.

INTRODUCTION

The purpose of this book is to make a comprehensive survey of Man and his development—past, present and future. The author has undertaken first of all to summarize the vast array of factual evidence already accumulated which may help us to determine what are man's specific potentialities, and what have been the varying conditions which have hitherto aided or hindered their progressive realization. He then discusses in scientific detail what the human race is now in a position to accomplish.

The objective of his investigation is not so much theoretical as practical. For this reason a brief preliminary explanation of the mode of approach seems desirable, if only to forestall a host of minor objections which would be quite irrelevant in the present context.

The immediate need of the scientist is to prove things, whereas that of the practical man is to manage situations. When we turn from the scientific study of the laws of nature to the study of the history of mankind, we find ourselves in an entirely different context. We seem to be watching a great drama—a drama of the modern type in which the spectators themselves play an active part. At almost every stage more than one alternative is possible. Which of the various alternatives is actually realized depends largely on ourselves. In the past it has depended, as often as not, on pure chance. Nature herself has proceeded by making the most of chance variations; and what Nature has done, slowly and blindly, man may now do with his eyes wide open and of a clear set purpose.

Today we need no longer fear that the scientist will step in to assure us that it is all in vain, since "what will be will be, and we have no choice." Even in the material world, every radioactive atom has two courses open to it—to continue as it is, or explosively to emit an alpha or a beta particle and add another step towards radioactive decay. And the quantum physicist maintains that "no force, either external or internal, known or unknown, can eliminate this element of indeterminacy—of choice or chance." One such change may cause the mutation of a gene; one such mutation may produce a new type of individual; one such individual may have it in his power to influence the course of history.

One of the most recent discoveries has been the detailed knowledge of the close relations between man's conscious processes and the physiological processes going on in his material brain. The behavioristic school, which developed on the Continent and in America, has resolutely sought to formulate all psychological conclusions in purely physical terms, banishing any type of exposition which might involve such concepts as consciousness, mind, choice or purpose. Accordingly, it will be, I imagine, not from biologists or geneticists, but rather from psychologists of the behavioristic school that the criticisms of the author's proposals will chiefly come. They will insist on strict physicalist interpretations and an assumption of strict determinism—sound principles from a methodological standpoint, but (so I would hold) quite untenable as rigid dogmas.

So long as our aims are restricted to securing cogent proofs for the abstract generalizations of science, such caution is admirable; but it is impossible to keep to it when we turn to the concrete problems of everyday life. "Even the most thoroughgoing determinist," says Dorothy Sayers in one of her stories, "will swear at the lab boy when he drops a test tube or turns the wrong tap": in spite of his theoretical scruples, he tacitly assumes that the boy *could* have acted otherwise. Thus, when we are concerned with practical affairs, it nearly always proves necessary to introduce a number of supplementary, and often purely tentative, assumptions, basing them so far as possible on an assessment of their relative probabilities. This is the approach that Robert Graham has adopted in writing *The Future of Man*. The logical positivist will doubtless feel tempted to challenge many of his statements on the ground that they can neither be tested nor checked. But that is inevitable in formulating any line of conduct. We could plan no campaigns; we could fight no battles;

we could not even carry on our daily business, unless we were prepared to face risks.

There are two essential steps in Graham's contentions which are especially likely to provoke objections. Let me therefore briefly indicate where they arise. First of all, the very attempt to discuss and decide on a course of action which shall aim at the betterment of the human race and the conditions under which it lives, implies tacit assumption that we *can* make a genuine choice, and that we *do* possess a set of values such as will furnish objective criteria for judging the goals at which we are to aim. If, as both Watsonian behaviorists and Freudian psychoanalysts declare, all our actions were mechanically determined, and our impressions of choice and free will mere baseless illusions, then it would be useless to strive against our fate.

This pessimistic fatalism is still widespread among many members of the educated public; and it is a mood that we must overcome if we are even to think of achieving what is proposed in *The Future of Man*. We may frankly recognize that the objectivity of values, the possibility of genuine choice, and the conceptions of purpose and free will are things that cannot be irrefutably proved by empirical evidence; they are assumptions rather than inferences, based on direct experience rather than on reason or argument. But they are assumptions which we must be ready to accept if we are to listen to the programs of action in the chapters that follow.

The second set of assumptions are likewise in conflict with the behavioristic psychology. We are reminded that, even in highly educated and civilized communities, human beings differ widely in their mental characteristics, and that their mental differences, like their physical differences, are the result partly of their material, cultural and social environments and partly of innate and hereditary tendencies. Unlike all other species, man is by nature endowed with the capacity to improvise his individual adaptations to the circumstances in which he happens to find himself. When lowlier mammals wandered into the chilly North, most of them failed to survive until at last by the cumulative effect of accidental variations a few families or stocks developed fur coats as part of their natural structure. Man, however, makes and dons an artificial overcoat. Before birds could fly, they had to wait until they had changed their forelimbs into wings. Man builds aeroplanes. And so with the dawn of civilization further evolution of man's innate abilities virtually ceased. Instead, a far more prompt and effective mode of progress came into operation —the conscious acquisition of knowledge and skill, and the transmission of the results, not by biological inheritance, but by social tradition. Not every member of the group invents the new idea for himself, but only some genius; and through imitation, education or verbal communication, the idea is passed on to many. Henceforth the new adaptations are stored, not in man's hereditary structure, but in his cultural environment. Of course, this does not mean that all genetic variation has ceased.

It is true that in democratic countries the chief, if not the sole, emphasis is popularly placed on the social and environmental factors. We *know* that the cultural environment exercises a powerful influence; and we *know* pretty well how that environment can be changed. On the other hand, even if we acknowledge that heredity continues to play a part, we do *not* know (so it is commonly said) whether its influence is really important, still less in what ways it acts. It is readily admitted that, as a rule, the less intelligent and the less self-restrained members of the population produce the larger families, and have done so for many years. Nevertheless, so we have been told, there is no conclusive evidence that this in any way affects the innate quality of the population as a whole.

The answer is that in practice such questions can hardly be decided on the narrow basis of what we already *know*. Once again we have to weigh probabilities. Nor is it sufficient to take into account only the hypotheses with the highest probability: we must also have an eye to the gravity of the different issues that may be at stake. Though the probability of a particular mishap is comparatively low, yet, if the effect would be both grave and irreparable (such as fire, or sudden death), the prudent man is ready to insure heavily against it; if, on the other hand, the effects would be relatively trivial (as the loss of an umbrella or a purse), then, even though the likelihood is much greater, we are willing to face the risk. Now a general decline in the innate mental endowment of the general population would indeed be a grave and irreparable result. Consequently, we must look far more closely into the basic problems.

When we do so—when we examine the relevant data chapter by chapter as Robert Graham sets them out—then, although I readily admit that the details are at best only a matter of probabilities, nevertheless their cumulative effects are impressive. Much of the evidence has come to light only during the last few years, and it still remains buried in technical papers published in inaccessible journals. As a result the popular notions about mental difference and human heredity—even among educated persons— are still largely erroneous, and almost wholly behind the times.

Let me, as a psychologist, briefly summarize the evidence so far as it is available. What we usually describe as mental characteristics are of two kinds: cognitive (or, as I should prefer to say, directive capacities, since they include practical as well as merely intellectual abilities) and conative or dynamic capacities (roughly what are known as emotional and moral characteristics). Among both we may distinguish those that are general and those that are specific. Thus on the "directive" side there appears to be an innate, general, cognitive ability, which Binet (following Spencer) called "intelligence"—giving the term a far more restricted meaning than is customary in popular parlance. A mass of converging evidence drawn from various sources strongly suggests that differences in this general capacity depend largely on the individual's genetic constitution.[1]

We now know that practically all innate tendencies are transmitted in accordance with Mendelian principles. From this it follows that not only the resemblances between members of the same family, but also the differences may be the effect of their genetic endowment—a point that is not commonly realized. Further, most graded characteristics—"intelligence," for example —are dependent, not on a single gene, but on a large number of genes. The pool of genes from which any particular individual draws his own assortment is made up partly of those carried by his father and partly of those carried by his mother. Thus a wide

---

1 "In mental ability . . . 65 to 80% of the differences are due to heredity."—L. O. Dunn and Theodosius Dobzhansky, *Heredity, Race and Society* (New York: New American Library, a Mentor book, 1952), p. 22.

diversity of combinations and re-combinations is conceivable, even for the children of a single parental pair. Consequently the range of innate differences in intelligence proves to be unexpectedly wide; they vary from that of the dullest idiot in a mental institution to that of a Leonardo or a Shakespeare.

An acquaintance with the principles governing biological transmission has enabled breeders of livestock, crops and garden flowers to develop entirely new varieties with qualities which no farmer or horticulturalist a century ago would have thought it possible to produce. To give but a single instance picked out almost at random: about 7 percent of all the milk cows in the U.S.A. are enrolled in one or other of the associations for the improvement of dairy herds established some 50 years ago. These cows, which have been scientifically bred, produce at the rate of about 9,700 lbs. of milk containing 380 lbs. of butterfat—roughly double the average rate and quality achieved at the end of last century.

We have every reason to believe that the genetic principles which apply to all other bisexual organisms apply also to man, and to man's intellectual and temperamental qualities as well as to his physical characteristics: for in the last resort the basis of such qualities is a physical or a physico-chemical structure. We must, however, remember that what are transmitted are not the qualities as such, but only the genes, that is, the tendencies or potentialities. Thus a man's genetic constitution must set a fixed upper limit to his intellectual achievements; but whether or not his full potentialities are actually realized must depend on conditions obtaining after he has been conceived—that is to say, on prenatal conditions during embryonic growth and on the postnatal or environmental conditions during the earlier years of independent life and development.

We can trace the way the genetic principles operate when we investigate the rise and fall of certain stocks or families, of social classes or groups, and of great civilizations of the past. In each case a plurality of concurrent causes has nearly always been responsible. Nevertheless, changes which tend to alter the genetic pool of the total stock often seem to have played the dominant role. Consider the intriguing problems raised by what historians have taught us to call the "centuries of genius." The first and the most remarkable was the century which lasted from the battle of Marathon to the death of Socrates (490-399 B.C.). During that period Athens, with a population of under 70,000 freeborn males of adult age, produced fourteen men of the highest intellectual eminence—a proportion of 1 in 5,000, whereas the number of geniuses of equal merit among other civilized nations averaged throughout the centuries less than one in several million.

To suggest explanations for the rise of dominant nations is comparatively easy for the speculative historian. Gibbon, musing among the ruins of ancient Rome, found it harder and more urgent to account for their decline and fall. Once a civilized society has established and organized itself, how does it ever come to pass that in the course of another brief period it so often fails in the competition for survival with one or other of its less civilized neighbours? Faced as they themselves were with this very problem in the course of their own lives, Plato and Aristotle both maintained that the success or failure of a state as a state depended first and foremost on the character of the men who govern it. Unfortunately, as Plato himself relates, the freedom, the security, the life of comfort and enjoyment which the vigour and efficiency of the rulers had step by step succeeded in gaining for all members of the state brought with them special dangers of their own. Security, comfort and enjoyment are most readily accessible in towns. But, unlike the country, urban conditions seldom make for the preservation of either physical health or lofty moral standards. The less intelligent inhabitants who were formerly poor and downtrodden now begin to multiply; and when all are free, the day-to-day government tends to pass into their hands simply because the less intelligent are the more numerous and so have a wider voice. Being less intelligent, they are more suggestible; and thus easily swayed by any sufficiently skillful and unscrupulous leader or a gang of selfish seekers after power—"demagogues," as Thucydides calls them, who promise the multitude the wages, the leisure and the entertainment they desire, and, thinking only of the present, care little for the future or for the state itself. And the remedy for all this? Plato's twofold prescription has a decidedly modern ring—selective breeding and improved education.

Similar changes can be discerned in the period preceding the

decline of each of the great civilizations of the past about which
we have reliable information. The story of imperial Rome affords
the most striking instance. Cicero, Tacitus and the satirists all tell
much the same story. Cicero significantly observes that at Rome
"the lowest of the citizen classes were called the proletariat, because their only contribution to the state was their *proles*" (their
numerous offspring). And Augustus introduced a series of laws
to check the diminishing numbers of the patrician class by im-
posing taxes on bachelors, forbidding senators to marry below a
certain rank and granting substantial privileges to those with
three or more children. All to no avail. Two things alone, says
Juvenal, would satisfy the dominant masses—*panem et circenses*,
food and the spectacular amusements of the circus; on these more
money was lavished than on either the army or the administration.

No doubt the Greek political philosophers oversimplified
both the conditions of their time and the policies they discussed.
And the conditions of our own day are infinitely more complex.
Nevertheless, two great thinkers, just because of their simplifying
insight, succeeded in picking out what are the essential ingredients of all such situations. One point Plato's discussion particularly makes clear. The various *external* causes—numerous,
insidious and obscure as they often are—constitute no more than
accessory or precipitating causes. However adverse such circumstances may seem, a nation that consists of a hardy, intelligent and
energetic population, guided by hardy, intelligent and energetic
leaders, "always seeking," as Aristotle puts it, "the good of the
community rather than themselves," will nearly always emerge
victorious from every struggle, social, military or economic. On
the other hand, when the health, the intelligence, the industry
and the self-discipline of the nation begin to wane, no amount of
legislation or education, no kind of social or political reform, is
likely to save it.

Robert Graham continues the story by pointing out how the
deliberate slaughter of most of the ablest members of the state
occurred during the French Revolution and has been repeated
with still more disastrous consequences in Russia, China and other
countries where masses predominate sufficiently to create for
Communists an opportunity to organize proletarian revolutions.

As a result of psychological inquiries undertaken during the last fifty years I could quote concrete evidence that a diminished proportion of the able is to be found in other countries also. In London, during the various surveys carried out with the aid of teachers and social workers, we took special note of the type of home or family from which each child came. We found that families which included a scholarship winner contained (on an average) only 2.3 children, whereas those which included one or more mental defectives contained 4.7—more than twice as many. A little calculation showed that, if we took such figures at their face value, they would imply a fall of about 1.9 I.Q. points in each successive generation. But of course a mere arm-chair deduction such as this is by no means conclusive. The processes of genetic transmission are far more complicated than a simple inference of this type tacitly assumes. On weighing the actual data from various sources, we came eventually to the conclusion that, in the course of a single generation, there has been in all probability a decline in the average level of the London population amounting to at least 1 point, but not more than 2 points. In the rural districts the decline usually appeared considerably larger. But it should be noted that a downward shift of only 1 point in the general mean would have a marked effect upon the numbers in the tails of distribution. If we take 70 I.Q. as a borderline for the subnormal, the proportion of such cases would be increased from 2.3 to 2.6 percent; while with a borderline of 130 I.Q. for scholarship winners, their numbers would be reduced from 2.3 to 1.9 percent. If this trend persisted through one generation after another, the effects would quickly mount up: in about three generations—less than a century—the number of scholarship winners of the same quality would be halved.

But it is not sufficient (as is common in discussing these issues) to confine attention to the effects of the differential birth-rate within our own nation or country, whatever that may be. We must also take note of the population trends in all the countries of the world. The gravest danger—and it is this that the author of *The Future of Man* wishes to emphasize above all else—is not just a decline in this or that civilized society, but a decline in the genetic endowment of the whole human race. We have already seen one such disaster, namely, that which occurred during the Dark Ages; and it took Europe centuries to recover from it. Accordingly, in the current discussions about the world population, let us bear in mind that the problem is one of quality as well as of quantity.

Admittedly these grim forebodings are speculative; yet one thing, I believe, we can affirm with absolute confidence. Whether or not the reader feels inclined to endorse the author's conclusions, he must surely agree that his book has made out the strongest possible case for focusing both public and scientific attention more sharply on the whole problem. What is more, he points out the ways in which deterioration (as he believes it to be) may be arrested, and what benefits would then accrue.

Certainly in our eagerness to improve the lot of our own generation, we tend to close our eyes to the possible effects of our various measures on the generations still to come. The overall efficiency of the individuals who make up a state or a nation must in the last resort depend upon the chromosomal pool. Improvements in material, educational and social conditions no doubt confer welcome advantages on those immediately affected; but by themselves they can ensure no permanent effects. On the other hand, alterations in a nation's genetic constitution are likely to prove irreversible. All over the world, the last half century has witnessed radical modifications in the traditional conditions that regulate the various mating systems, and in the social incidence of human fertility. The expansion of transport facilities, the increase in individual freedom, the new modes of production, the changes in the educational ladder, and above all the progressive reduction in the barriers which separate different social classes, different economic groups and different races—all these are visibly transforming the characteristics of the various human breeds. Almost inevitably they must be affecting the genetic constitution of what have hitherto been the dominant nations and the dominant stocks within those nations. Our growing knowledge of the physiology of the reproductive processes—leading to novel methods of birth control—are bound to complicate the problem yet further. How far they may operate for good or for ill we still cannot say. Hence the two crying needs of the moment are first a more resolute organization of scientific research, and secondly a general awakening of interest in the far-reaching issues which are involved. It is for these reasons that I should like to recommend, as strongly as I can, the considerations that the author has adduced, to the earnest attention of the intelligent public.

Professor Sir Cyril Burt, F.B.A.,
D.Litt., Ll.D., D.Sc.
University College, London

Part I

YESTERDAY

CHAPTER I

BEFORE MAN

More than four thousand million years ago the earth lay raw and without life. There were rocks and sand, sunshine and winds, rain, streams and seas. Yet in them or on them there was nothing alive—no trees or grass, no moving creatures—only desolation. Then certain very small but immensely significant events took place: the first living things which could survive and reproduce themselves floated upon the waters of the sea.[1] For millions of years the descendants of these infinitesimal blobs of living jelly drifted aimlessly with the warm currents. During all of that time they were growing and dividing and in this way multiplying themselves.

When the first of these primordial populations had multiplied until the available food and space no longer could sustain all of it, there began among its members a grim contest to survive. This elemental competition spread throughout the whole organic world and was intensified by conflict with other types of living things. It has since shaped every form of life on earth. What gave direction to the lethal contest was the fact that occasionally one or more of the contestants would differ in form or habit from its parent. A few of the differences gave to their

----

1 Here we are concerned not so much with beginnings as with outcomes. However, for learned conjectures about the beginnings of life see one or more of the following:

Hermann J. Muller, "Life," *Science*, Jan. 7, 1955.

George W. Beadle, *Physical and Chemical Basis of Inheritance* (Eugene: Oregon State System of Higher Education, 1957), p. 9.

A. I. Oparin, *The Origin of Life* (New York: Dover Publications, 1938), especially pp. 248-250.

Linus Pauling and Harvey A. Itano, editors, *Molecular Structure and Biological Specificity*, Publication No. 2 (Washington, D. C.: American Institute of Biological Sciences, 1957).

F. Clark and R. L. M. Synge, editors, *The Origin of Life on the Earth* (New York: Pergamon Press, 1959).

possessors some advantage over their brethren. These differences enabled the vantaged ones more surely to survive and multiply their particular kind, despite inhospitable surroundings.[2] Subsequently others would occur with some advantage over the best previous types. These then would thrive and increase, often at the expense of the earlier kinds.[3]

It was the struggle to stay alive and the resulting natural

––––––––

2 ". . . all organic beings, without exception, tend to increase at so high a ratio, that . . . not even the whole surface of the land or the whole ocean, would hold the progeny of a single pair after a certain number of generations. The inevitable result is an ever-recurrent Struggle for Existence. It has truly been said that all nature is at war; the strongest ultimately prevail, the weakest fail; and we well know that myriads of forms have disappeared from the face of the earth. . . . This preservation, during the battle for life, of varieties which possess any advantage in structure, constitution, or instinct, I have called Natural Selection; . . ." — Charles Robert Darwin, *Variation of Animals and Plants Under Domestication* (New York: Orange Judd, 1868), pp. 5-6.

3 The earliest creatures reproduced by simply dividing in two. Each half then became a new individual, able to grow and re-divide again. Although the majority of these creatures might be consumed by enemies, the death of the individual is not inevitable under this system. For example, the whole population of amoebas today quite possibly consists of subdivisions of the first amoeba.

Later, but still far back in the evolution of life, a different way to reproduce began to manifest itself. This required the union of germinal substance from two separate individuals. Its great advantage arose from the fact that the offspring, instead of almost always being exact duplicates of a parent, were never exactly like either parent. Sex is nature's greatest invention for speeding the rate at which a species can change and thereby make the most of its environmental opportunities.

". . . sex is Nature's method of shuffling the cards, so that the offspring can play the game of life with a different hand from that dealt to their parents, and sometimes, though not always, a far better one. Since the game of life is an elimination contest, the living beings dealt a better hand have lived on and the others have not. Thus sex as we experience it is far more than an act of pleasure or even of procreation. It is a reshuffling of life's cards, out of which can come greater success." —John Langdon-Davies, *The Seeds of Life* (New York: New American Library, 1955), p. xv.

So great is this advantage that the sexual reproducers have reached an incomparably higher state of development than any of the simple fissioners.

As living matter grew more complex, a second sweepingly advantageous principle came into play. It utilizes the fact that, once the offspring can fend for themselves, it is unnecessary for the parents to

selection which made the occurrence of life, originally only a self-duplicating molecular linkage, so portentous an event. Through the working of these forces during billions of years there slowly evolved all the countless wonders of the organic world, including the very eyes with which you read these words and the brain with which you comprehend them.

Vastly varied are the present forms of life which have developed under natural selection. Far more vast is the number of forms which throve for a time but later died out as conditions changed or as new enemies or more successful competitors came into being.[4]

Sometime during the early development of many-celled animals there began to appear in their tissues a type of cell which could conduct impulses from one part of the body to another.[5] This beginning of a coordinating system, never accomplished by the plants,[6] opened the path which led eventually to man, the creature endowed with the most highly developed of all neural systems. The beginning of conducting cells can be called the beginning of intelligence.

---

continue to live. In fact, their demise leaves more food and space for the young. Thus, there is opportunity for more descendants, hence for more variations for natural selection to choose from and consequently more rapid and perfect adjustment of the type to the exigencies of existence. The biologic advantage of eliminating the older generations is very great, and death of the individual is an invariable characteristic of all multicellular species. In them only reproduction enables life to prevail over death.

This, then, we know with uttermost certainty: the flesh will wither and fall away, but in the time of its fullness it may renew itself in offspring and so have everlasting embodiment upon the earth. We can regard more calmly the fact that time consumes us when we see it as a contribution to the well-being of succeeding generations—and even more calmly if we ourselves have contributed new lives abundantly to the next generation, for whom we sacrifice in life and for whom, eventually, we make the ultimate sacrifice so that the part of us which lives on in them may live the better.

4 ". . . during the time the earth has been inhabited, something like 95 or 98 percent of all species have become extinct." —Thayer Scudder, *The Next Ninety Years* (Pasadena: California Institute of Technology, 1967), p. 184.

5 Jellyfish are among the simplest creatures to possess specialized conduction (i.e. nerve) cells.

6 Venus's-flytraps and sensitive peas may be exceptions to this, but **not very significant ones.**

CHAPTER II

THE HUNTED TURNS HUNTER

The tree-dwelling ancestors of man were among the most intelligent beings of their distant time. When these creatures finally abandoned the trees and walked upright, freeing their hands for use as implements of the mind,[1] there began the most successful evolutionary drive toward higher intelligence ever witnessed in nature.

As ground-dwellers, these creatures came within reach of the great beasts of prey. By day they were hunted down; at night they were attacked by stealth as they huddled together. They could not compete in strength, ferocity or speed with the other large denizens of the earth. Armed with little except hands and what their superior brains enabled them to make with hands, they had to think or die.[2] For untold thousands of years most of them died violent deaths early in life. Only a few in each generation had the good fortune and the ability to out-wit their enemies and out-think their prey and so survived long enough to beget and rear young.[3] Maladroit or stupid hunters lost their lives and, if they had offspring, left most of them to starve or be devoured.

---

1 ". . . only man achieved a completely bipedal gait, with the primate hands completely 'emancipated' from walking." —Weston La Barre, *The Human Animal* (Chicago: University of Chicago Press, 1954), p. 73.

2 "To avoid enemies or to attack them with success, to capture wild animals, and to fashion weapons, requires the aid of the higher mental faculties. . . . These various faculties will thus have been continually put to the test and Selected." —Charles Darwin, *The Descent of Man and Selection in Relation to Sex* (New York: D. Appleton, 1871), p. 873.

3 "The growth of the brain . . . is simply the result of the fact that Primates were able to maintain themselves in the struggle for existence by the exercise of wile and cunning. . . . The wile and cunning of the earlier Primates have become the intelligence of the higher Primates, and Man himself has surpassed all other members of the animal kingdom in his capacity for mental activities of the most elaborate kind." —Sir W. LeGros Clark, *History of the Primates*, 3rd ed. (London: Printed by order of the Trustees of the British Museum, 1953), pp. 179-80.

Natural selection was operating on man with grimmest intensity. Perhaps no other creature which survives today has undergone so severe and protracted a period of selective elimination. Yet here and there small groups managed to survive because they had the intelligence to employ their newly-freed hands more effectively than did the others. They used sticks, stones and clubs well enough to defend themselves and to secure sufficient food for survival.[4] In each generation it was the more effective ones who succeeded in rearing most of the next generation.[5]

The steps in the development of man's brain are revealed by the progressively larger brain-cases which appeared with the passage of centuries. Basing our judgment on the improvements in tools and weapons which took place as brains became bigger, we employ reasoned conjecture in the coming pages to reconstruct some of the ways in which natural selection worked to bring about a doubling in size of the brain of man.[6]

NATURAL SELECTION AND THE BRAIN OF MAN

Early Implements

Many edible nuts are too hard for even a caveman to crack between his teeth. They were, therefore, useless to early man until some genius of his day discovered that any nut could be opened if it were placed upon one stone and struck with another. The family of this originator, being thereafter better fed, in-

---

4 Crude and puny though sticks, stones and clubs were, they were *weapons* and their possessors were the first creatures which could kill without having to come in contact with their antagonists.

5 "As the great beasts grew larger and either faster or more formidable, man became ever more watchful, ever more successful in pitting his wits against mass and power, more and more adept at slipping out of trouble, and as the challenge grew greater, so did his brain, for the laggards on both sides got left behind in the race for the future." —Norman J. Berrill, *Man's Emerging Mind* (New York: Dodd, Mead, 1955), p. 109.

6 "The most astounding phenomenon of Human evolution is the rapid increase in brain size during the Pleistocene. . . ." —Ernst Mayr, *Animal Species and Evolution* (Cambridge: Belknap Press of Harvard, 1963), p. 650.

". . . the most marked phylogenetic trend in the evolution of man has been the special development of the brain, and . . . the characteristic human plasticity of mental traits seems to be associated with the exceptionally large brain size." —M. F. Ashley Montagu, *Anthropology and Human Nature* (Boston: P. Sargent, 1957), p. 119.

creased and throve exceptionally. The superior thinking ability of this early man was inherited by several of his children, and so was multiplied and spread to many descendants.

Perhaps centuries later, a man sat cracking nuts with a stone. Suddenly the stone broke and one of its broken edges cut his hand. Where prior men in the same situation had only broken stones and injured hands, the clearer mind of this man told him that he held a useful thing. The edge had cut through his skin and drawn blood; therefore it might cut through the skins of the small animals he caught, making the flesh easier to get at. He and his kind and those intelligent enough to imitate them flourished thereafter and increased in number, for they had a cutting tool which made meat-eating quicker and gave them more time for hunting. It was the first crude knife. Many of the descendants of this exceptional man became increasingly skillful at breaking and chipping hard stones into sharper tools and weapons. Because of this they were better able to kill and consume meat and to defend themselves from attack. In this way, the most skillful and intelligent among them continually held the advantage in the struggle to live and rear their children. Natural selection favoring better knife makers went on for hundreds of thousands of years.

A great many centuries later, it may well be, a young father, foraging for his brood, came upon a long, stout stick, splintered at one end. He pulled and chewed at the lesser splinters until only the main point was left. It seemed to him a very good point, for it was sharper than the digging sticks which the women used. In his mind remained the memory of a fearful night during boyhood when a great cat had charged his huddled family and dragged away one of his little sisters, despite frantic clubbing by the whole family. Now that he had small children of his own, his dread of such attacks was intensified. He had seen fresh panther tracks lately and another family had been attacked and some of them killed. Sharpened by fear, his glimmering intelligence told him that his pointed stick might be a better weapon against cats than the clubs which he and the other men carried. For many days, he kept the long stick near him, even when it brought jeers from others who saw him carrying what they regarded as a woman's tool.

Then one night there was a faint rustling close by. He

whispered a quick warning to his family. With that a dim shape charged at him in the darkness. Kneeling, he managed only to lift the point of his long stick toward the beast as it sprang. It clawed at him with frightful frenzy. But instead of seizing him, it fled! He had felt the creature strike the lifted point so hard that the blunt end of the stick was thrust backward into the earth.

Next morning, following a spoor of blood, they found a panther, dead from a punctured chest. The man with the long, sharp stick had kept himself alive where others without his fore-thought had been killed.

From that time, he and his sons and their sons carried im-paling-sticks whenever cats were near. Foresight had given them an improved means of defense. As a result more of them sur-vived than of the others, especially the others who did not or could not learn about impaling. Some of the offspring inherited the superior intelligence of the man who had first envisioned new possibilities in a long, sharp stick. As their families multiplied and spread, replacing families which were easier victims of cats, the intelligence of the humans in that area was increased.

Perhaps many generations later, one of the brightest de-scendants of the first cat-impaler mated with the daughter of an-other intelligent man. This other man had thriven in a neighbor-ing valley because he had thought to throw his club at fruits, nuts and small animals on the lower branches of trees. Now and then this brought down an extra meal. The technique had helped him to sustain his family during times of terrible scarcity, when many others had starved.

The man who knew how to defend his family from the great cats soon learned from his woman the new way to reach additional food. Their young family, better defended and better fed, grew in size and throve beyond all others in the region. Some of their children, with good mental inheritance from both sides of the family, showed an even higher order of intelligence than either of their parents.

Thereafter, the people in the old cat-impaling tribe and those in the club-throwing families were inferior in thinking ability to some of the scions who sprang from the fortunate union of the two lines. In a time of famine when battles over food were fierce these scions thought to cast their impaling-sticks, as well as

PRIMITIVE MEN USE GUILE TO TRAP A WITLESS
MASTODON AND STONE IT TO DEATH
(From a painting by Andre Durenceau in *National Geographic*,
December, 1955, p. 792.)

their clubs, at members of the other tribes. In this way perhaps the throwing of spears began. Its effect was deadly. Before long the older tribes, formerly the most intelligent and dominant in their regions, were replaced by others who were more intelligent and better armed.

With spears added to his clubs and cutting stones, man no longer had to be so furtive a food-gatherer. The hunted one gradually evolved into a bolder hunter. Little gains in producing sharper spears or keener stone knives gave to their devisers the advantage of survival during critical times. But the most telling gains were in increased sharpness of the minds which made the improvements in tools possible.

In some such ways did the selections of nature work. Thousands of times a few who had enough intelligence managed to survive until they could pass on their thinking ability to their offspring. Yet almost always the less favored in each generation were eliminated, usually early in life.

"However incomplete our knowledge of human ancestry, there is scarcely any doubt that the development of brain power, of intelligence, was the decisive force in the evolutionary process which culminated in the appearance of the species to which we belong. Natural selection has brought about the evolutionary trends towards increasing brain power because brain power confers enormous adaptive advantages on its possessors. It is obviously brain power, not body power, which makes man by far the most successful biological species which living matter has produced."[7]

"The one tremendous difference . . . is in brains. Mainly it was the series of additional brain mutations on the human branch of the primate tree (plus improvements in the hands and skeletal framework) that carried men so far beyond all other creatures, and left the apes and monkeys almost back where they were millions of years ago."[8]

---

7 Theodosius G. Dobzhansky, *Evolution, Genetics and Man* (New York: Wiley & Sons, 1955), p. 334.

8 Amram Scheinfeld, *The Basic Facts of Human Heredity* (New York: Washington Square Press, 1961), p. 204.

Speech

As the brain of incipient man continued to evolve, the jab-
berings and outcries of his fellows came to have increased meaning for him. At first he had but a few utterances to warn against
enemies. Then other sounds were added to call his family to food
or to the kill. Eventually leaders were able by speech and gesture
to direct their fellows in attacks on large beasts or scouts could
convey by words where to find a group of grazing animals that
had been sighted. Thus, slowly, language took shape.[9] Now each
person could understand his fellows better and join with them at
times to stand off attacks or to slay animals which none alone
could overcome. Language is the greatest of all man's early accomplishments. His tools were crude and his brain was still far
from equalling that of his descendants, but with the development
of language he could collaborate more effectively and transmit
more information to his fellows and to his children. Some animals
have a variety of emotional cries; man alone has articulate speech.
The dawn of true humankind was beginning.

"A word is an instrument of thought. . . . The naming of
things is the great difference that separates the human mind from
animal minds. . . . Man's brain evolved to the stage where he was
able not only to think of things (through images or whatever
animals use) but also to see the value of using names for them.
With language the human species entered a new world, began to
think more clearly, became capable of reflection on the past and
penetrating the future. . . ."[10]

By means of language the new ideas of the brightest individuals spread so that many who never originated ideas could
now take advantage of the ideas of others.[11] As words gave wings

------

9 "This of course implies antecedent changes of structure notably
in the bones, cartilages and muscles of the mouth and its adnexa and in
the so-called 'speech-areas' of the brain cortex." —Notation by Sir Cyril
Burt in typescript of *The Future of Man.*

10 Norman L. Munn, "The Evolution of Mind," *Scientific American,* June, 1957, pp. 140-150.

11 "Language . . . connects mind with mind, making common
property of worthwhile thoughts. . . . The artifice to which we give the
name of language . . . was built up at a very early age. . . . Through it
the mind was stimulated to further activity and the cerebral cortex must
have developed accordingly." —G. G. MacCurdy, *The Coming of Man*
(New York: The University Society, Inc., 1935), p. 123.

to thoughts, the thinking of a few became useful to all who could understand and imitate. This was to permit, millennia later, a profound change in the character of the average man.

At this stage, early man was still as much hunted as hunter. He was but one of the many larger animals and, when barehanded, the most defenseless of them all. The great usefulness of speech, tools and weapons gave the advantage to those with the sense to use them best. As the brain of man developed under this type of natural selection, early man came to stand out more and more distinctly from all other living creatures.

Fire

As we can surmise how the early crude implements were discovered or produced, we can also see how fire might have been captured and utilized by later man. The first time could have been a cloudy, gusty night, hundreds of thousands of years ago, when a great pine was set afire by a bolt of lightning. Nearby a headman and his family cowered in their cave, awed by the violence of nature. To most of them it seemed a sign of the anger of the gods.

When day broke, a part of the pine which had been dead and dry was still afire. The humans ventured closer, staring at the flames and smoke which rose from the once familiar tree.

An old woman called out that the smoke was the ascending spirit of the dying tree. The group regarded her with great veneration for her understanding of this strange occurrence. But the headman—who had destroyed small trees himself—scorned this belief. He knew that trees died differently. He watched intently as flames ignited the very limbs up which he used to climb to look across the hills for game. Now and again a burning branch would fall to the ground and set aflame, for a few moments, the pine needles and twigs on which it fell. As he watched, in his mind there grew a dim comprehension of the workings of fire. Driven by curiosity, fortified by his new understanding, he seized a dry branch from the ground and held it in a flame. The others shrieked warnings that this was defiance of the tree-god. The old woman indignantly struck the branch from his hand but he bowled the meddler over and was not interfered with again. Back into the flame went the branch. The man's fingers scorched with the heat. The hair on his outstretched hand was singed, but the fire licked at the dry branch and set it aflame. With a shout, he held up his burning branch. It flickered out, but he thrust it back into the fire and it was relighted more brightly. Bearing the first torch, he strode to his shelter, the group trailing after, and there laid it upon the ground. Quickly, he put dry needles and twigs beside it. The flames crept through the fuel, as he had hoped they might, and on that day man first had and kept fire. It warmed him in the cold. At night it drew animals. He could see their eyes gleam as he looked out from his cave. But no longer did they threaten him in his shelter.

Often, at first, he burned himself while tending the fire. Later, seeing how a breeze made the fire burn brighter, he learned to blow upon it when it smoldered low. Thus he revived his captured servant.

Others, hearing of the new wonder, came to see. The old woman, never one to let truth spoil a good story, exclaimed to all that this was a gift from the gods. Gathering faggots and kindling them, some of the awed visitors carried brands to start fires before their own caves. Thus the use of fire spread, warming and protecting its possessors. This was man's first conquest of the forces of nature, the first source of power outside his own muscles. With fire he left the endless procession of ordinary creatures and became the first super-animal on earth.

A few grasp an idea quickly, but most take a long time to follow after. This was especially true in those times, for man was not yet as intelligent as he was to become. Fear impeded his progress. To most of the cave folk, fire was a frightful thing. Many avoided it, fearing the displeasure of the spirits if they used it. In other places, fire-worshipping cults sprang up and spread, limiting fire to holy uses. It appears to have been many thousands of years after the capture of fire before it was ever used to cook food.[12]

----

12 The earliest evidence for the use of fire comes from a site which is about a million years old.

The earliest known cooking was done about 43,000 years ago. It seems incomprehensible to us that man should have had fire for thousands of years before he ever came to use it for cooking. It may seem equally incomprehensible to future generations that man learned how to improve plants and animals thousands of years before he began to use this knowledge so as to improve himself.

For generations the descendants of the first fire-master
throve exceptionally and thus the brains and courage of this
genius were multiplied. In fact, so great was the advantage of fire
that no type of man who did not understand and control it could
long compete against those who did.[13] Today no tribe of man
remains which does not have fire.[14]

---

13 ". . . the use of fire must at first have spread most rapidly among
the more far-sighted, energetic, and determined members of some groups
which lived near the coldward border of the region then inhabited by
man. We may reasonably picture a stage when the sparse population of
certain regions of that sort consisted of two types. One comprised relatively competent people who had learned to use fire and were thereby
stimulated to make new inventions, including in due time those con-
nected with cooking. The other was a less competent group who did not
think the profit to be gained from the fire was sufficient to pay for the
labor involved.

"Other things being equal two such groups are bound to increase
at different rates. The new invention helps to preserve the lives of its
users. It saves them from contracting diseases which are fostered by exposure to wet and cold. It helps to ward off wild animals which might
kill little children. It encourages inventions, such as the spear with a
point shaped in the fire, and especially the art of cooking. Such inventions give their users the great advantage of a food supply more abundant
and perhaps better than that of their neighbors. The result must be a
higher rate of survival among their children than among those of the
non-users.

"Another factor enables the fire-users to increase more rapidly than
the others. With the help of fire they are able little by little to spread
into cooler climates where the non-users do not follow because they
cannot there be comfortable. Thus selective migration occurs. In the
newly occupied regions, people of the more intelligent and competent
type can intermarry only with one another, whereas their former com-
rades in the old home intermarry not only with one another but with
the non-users of fire. Thus the intelligent fire-using type becomes established in the newly occupied regions and the culture of those regions becomes higher than that of the warmer regions. The center of cultural
progress and the optimum climate both shift from warmer to cooler
regions. Cool weather no longer has its former ill effect on comfort and
efficiency. On the other hand, in the new region the ill effects of undue
heat are lessened because the hot period is not as long as in the warmer
climate. In due time, to be sure, the biological advantage of the new
region may diminish through influx of the less competent type. This will
happen if overreproduction, war, pestilence, climatic change, or other difficulties drive the less progressive type out of its warm home or if the
means of making and using fire become so well established that it is easy
for incompetent people to use them.

"We may sum the matter up by saying that the presumable effect

Observations

In those remote times there were men of many patterns—far more varied even than today. Ape-men, small-brained men and giants, all dwelt in their chosen areas. At times they met and battled to the death. Slowly the less intelligent competitors in each area were exterminated by those who were able to devise better weapons, or could use them more effectively.

The threat of hunger hung like a dark cloud over all people who lived by hunting. There were times without number when there was not food enough for all. Only the keenest and most foresighted, who could find or capture enough nourishment to stay alive, survived to repeople the wilderness when food became more abundant.[15] The dull-witted and the uncooperative went hungry and left fewer progeny than the others or, more often, they and their children starved. Only those who left children had any chance of imparting their qualities to the next generation.

The struggle between men for survival did not consist solely of open battles between man and man. More often it consisted of

---

of the invention of fire was (1) to create a division of population into fire-users and non-fire-users; (2) to stimulate other inventions among the fire-users; (3) to cause their numbers to increase faster than those of the non-users; (4) to enable part of the fire-users to migrate into regions previously unoccupied because too cool and there to increase rapidly; (5) to shift both the optimum climate and the center of progress into regions cooler than the previous optimum; . . ."   —Ellsworth Huntington, *Mainsprings of Civilization* (New York: Wiley, 1944), pp. 403-4.

14 The Andaman islanders still do not understand how to kindle fire and so must tend it perpetually, through rain and storm, as in the beginning. —C. S. Coon, *Story of Man* (New York: Knopf, 1954), p. 61. Also "Living Stone Age Tribe," *Science News Letter*, Sept. 15, 1956, Vol. 70, No. 11, p. 166.

15 Experts have estimated that, during the thousands of centuries when man was increasing most rapidly in intelligence, only a few would remain alive during a severe winter.

". . . two eminent authorities, Professor Herbert J. Fleure and Grahame Clark, made independent estimates of the population of Britain in Upper Paleolithic times under a subglacial climate. They believe that it was in the range of 250 to 2,000 individuals and that it might have fallen to approximately 250 during rigorous winters." —Nathaniel Weyl and Stefan T. Possony, *The Geography of Intellect* (Chicago: H. Regnery, 1963), p. 80.

one family, or group of families, indirectly reducing the number of competing families. For example, when there was not food enough for all—and in temperate areas this was the prevailing condition in winter—those who captured the existing food deprived the others of it. Thus, a very able hunter might starve competing families without ever coming face-to-face with them.

And so through thousands upon thousands of years, some of the more intelligent and resourceful members of mankind managed to survive and to replace those who were less intelligent. Now and then mutations in the direction of more brain cells and greater intelligence would occur,[16] giving their possessors advantage over other humans. As a consequence, they and their children who inherited these advantageous mutations would tend to survive and replace the less favored humans about them. At other times the mating of two exceptional people would produce a strain superior to the others. By such steps (favorable mutations and sexual recombinations) did man progress. Natural selection for intelligence was crude, slow and cruel, but it was salutary. In the course of a million years or so, men of a very high order of intelligence had appeared upon the earth.

The fact that man, naturally the most defenseless of the larger creatures, survived at all is a tribute to the intelligence his forebears possessed as they came down from the trees.[17] That this arboreal fugitive was able to transform himself gradually from relative defenselessness into the deadliest of all hunters is a tribute to the potency of his increased intelligence.[18] It made possible the most amazing success story in all of life and nature. Lacking horns, his brain and hands enabled him to develop spears

_____

16 "The human brain has doubled in size twice since man first came to reason and such physical changes, the result of mutations, are likely to occur again." —George R. Harrison, *What Man May Be* (New York: Morrow, 1956), p. 263.

17 "Man is neither particularly strong in body nor particularly agile in movement. If it were not for his brain he would be a rather pitiable misfit in most environments, and would probably have become extinct long before now." —Theodosius Dobzhansky, *The Biological Basis of Human Freedom* (New York: Columbia University Press, 1960), p. 104.

18 And, of course, cooperativeness and other important human attributes which are not the subject of this study and perhaps not so heritable.

which were more effective than any horns; without claws or fangs, he developed knives which were superior to either; given a poor sense of smell, he learned to enlist keen-scenting dogs as his allies. Through such hard-won steps the thinking creature, once so hunted by fierce animals on every side, came to slay his enemies and clothe his nakedness in their very skins.

"Man in the rudest state in which he now exists is the most dominant animal that has ever appeared on this earth. He has spread more widely than any other highly organized form; and all others have yielded before him. He manifestly owes this immense superiority to his intellectual faculties, to his social habits, which lead him to aid and defend his fellows, and to his corporeal structure. The supreme importance of these characters has been proved by the final arbitrament of the battle for life. Through his powers of intellect, articulate language has been evolved; and on this his wonderful advancement has mainly depended. . . . He has invented and is able to use various weapons, tools, traps, etc., with which he defends himself, kills or catches prey, and otherwise obtains food. He has made rafts or canoes for fishing or crossing over to neighboring fertile islands. He has discovered the art of making fire, by which hard and stringy roots can be rendered digestible, and poisonous roots or herbs innocuous. This discovery of fire, probably the greatest ever made by man, excepting language, dates from before the dawn of history. These several inventions, by which man in the rudest state has become so pre-eminent, are the direct results of the development of his powers of observation, memory, curiosity, imagination, and reason."[19]

"Of the high importance of the intellectual faculties there can be no doubt, for man mainly owes to them his predominant position in the world. We can see, that in the rudest state of society, the individuals who were the most sagacious, who invented and used the best weapons or traps, and who were best able to defend themselves, would rear the greatest number of offspring. The tribes, which included the largest number of men

---

19 Charles Darwin, *The Descent of Man* (New York: Modern Library ed. Reprinted from 6th ed., 1872), pp. 431-32.

thus endowed, would increase in number and supplant other tribes."[20]

It was your forebears and mine who, as a result of their clearer thinking, managed to live until they could transmit the magic of their intelligence to their children and thence on to us. In times of crisis it was those few in each generation whose intelligence shone a little brighter than the rest; who saw their way more clearly and followed it more successfully; who handed on the torch for us when almost all with lesser endowment fell early in the savage contest. Many were called into being and few chosen.

The selecting for intelligence went on for innumerable generations. In all that time, if there had been even one member of your ancestry or mine who failed to live to young adulthood or, living, failed to bring into being more of his or her kind, we never would have known life. The wonderment of our being is a surpassing miracle.

We know well how severe was the winnowing throughout the million or so years when creatures properly called man existed by hunting, for only a comparative handful of humans survived in any generation.

"Probably the people lived in small groups, as hunting and food-gathering seldom provide enough food for large groups of people."[21]

"At the time of Aurignacian man (i.e., Cro-Magnon cavemen who lived in Europe 30,000 to 40,000 years ago), there were perhaps a dozen hordes, each a few hundred strong, wandering in the whole area of France, and such hordes must have regarded it as a deeply impressive and puzzling event when they became aware that fellow men existed. Can we imagine even in the least degree what it was to live in a world almost empty of men . . .?"[22]

For more than a million years of hunting, natural selection kept the whole human population below one million individuals. Later, when man's discovery of agriculture made food much

---

20 *Ibid.*, p. 497.

21 Robert J. Braidwood, *Prehistoric Man* (Chicago: Natural History Museum, 1959), p. 68.

22 Oswald Spengler, *Decline of the West*, Vol. II (New York: Knopf, 1928), p. 34.

more abundant, so that the proportion of those who did not
starve to death or die fighting was greatly reduced, man was able
to increase in only 13,000 years from less than one million to
more than three thousand million. This tremendous upsurge indicates how sternly natural selection had kept killing off all but a
few of the most able during the thousands of centuries of the
hunting stage.[23]

Other factors, including strength and luck, entered in, but
in the human tribe the scales of advantage mostly tipped toward
those who were more intelligent.[24] As the ancient record is un-
earthed it shows, time after time, that larger-brained and more
intelligent forms of man appeared and replaced smaller-brained
types who made cruder weapons.[25] Intelligence came to be the
most human thing about human beings.

———————

23 "This observation is valid because the reproduction rate did
not increase at the time agriculture began." —Notation by Hermann J.
Muller in typescript of *The Future of Man.*

24 In the large apes, who also became principally ground-dwelling
and partially erect, the main evolutionary drive was in the direction of
greater strength instead of greater intelligence. Their muscular power
became prodigious. (See Figure 3.)

"Tests on immature gorillas have shown a muscular strength two
or three times that of a strong man. The strength of the adult male gorilla
is not exactly known because of his . . . active lack of cooperation in
the administration of such tests." —Weston La Barre, *The Human Animal, op. cit.,* p. 65.

However, gorillas are now undergoing extinction by tribes of men
who, although much smaller and weaker, have evolved a higher order
of intelligence and, with it, deadlier weapons than brawn and fangs.

"The gorilla alpha chain and the human alpha chain differ in two
residues, and the gorilla beta chain and human beta chain differ in one;
the average, 1.5, indicates that about 11 million years have gone by since
the derivation of these chains from their common chain ancestor—that is,
that the evolutionary lines leading to the present-day gorillas and present-
day human beings separated from one another about 11 million years ago.
The estimates made by paleontologists for this epoch range from 10
million to 35 million years." —Linus Pauling, "Molecular Disease and
Evolution," *Proceedings of the Rudolf Virchow Medical Society in the
City of New York,* Vol. 21, 1963, p. 341.

25 "In a general way we can say that the brain volume doubled
during the ten million years or so of man-ape evolution . . . and that
it has on the average doubled again during the last million years." —
Norman J. Berrill, *Man's Emerging Mind, op. cit.,* p. 70. (See Figure 4.)

"The gradual increase in the size of the brain . . . accompanied the
trend toward greater intelligence, the most important feature of human

Homo Sapiens Appears and Crowns the Hunting Stage

Probably in some part of Mesopotamia (then a garden of Eden), there emerged a type of man designated today as Homo sapiens (Man, the knowing one).[26] He appeared some fifty thousands of years ago and was the most intelligent type of human yet to evolve.[27]

In the course of centuries, as bands of Homo sapiens slowly spread, all other branches of the human tribe died out.

"The rest of the human story is that of Homo sapiens alone. No other kind seems to have lived long beyond that same point, about 37,000 years ago, at which modern men were coming into Europe."[28]

More intelligent types overran and exterminated the less advantaged, time and again, even as the Bantu are exterminating the Bushmen in Africa today.[29]

The sapiens form of man became a wonderfully successful creature. His descendants migrated to and dominated all livable parts of Earth. Some types of this new kind of man progressed, some retrogressed, depending on their nature, surroundings and the extent to which they absorbed more primitive peoples[30] but everywhere they predominated over all other orders of man and

---

evolution." —Clarence W. Young and G. Ledyard Stebbins, *The Human Organism and the World of Life* (New York: Harpers, 1938), p. 843.

26 "The cradling place of mankind has been a matter of much speculation. It is now generally accepted that this was located not in Europe, but rather in the extensive central regions of the Asian continent. From such a birthplace the race has gone forth in successive waves of migration to the north, to the east, to the south and to the west." — Frederick Tilney, *The Brain From Ape to Man*, Vol. II (New York: Hoeber, Inc., 1928), p. 767.

27 "The appearance of this new species marked the latest great advance in structural evolution and ushered in the age of modern man. Homo sapiens possessed a larger and better balanced brain than any of his predecessors. He was distinguished also by numerous refinements in bodily form and by superior manipulative skill and language capacity."—Carl J. Warden, *Evolution of Human Behavior* (New York: Macmillan, 1932), p. 137.

28 William White Howells, *Mankind in the Making* (Garden City: Doubleday, 1959), p. 259.

29 See Philip V. Tobias, "Physique of a Desert Folk," *Natural History*, Vol. 70, Feb., 1961, pp. 18-21.

30 Insofar as interbreeding was possible, there was probably some absorption of less successful peoples, through capture of their women.

Fig. 3. LARGE MALE GORILLA KILLED IN THE AFRICAN CONGO
Note the gorilla's greater size and strength, the man's larger brain case.
(Courtesy of the American Museum of Natural History.)

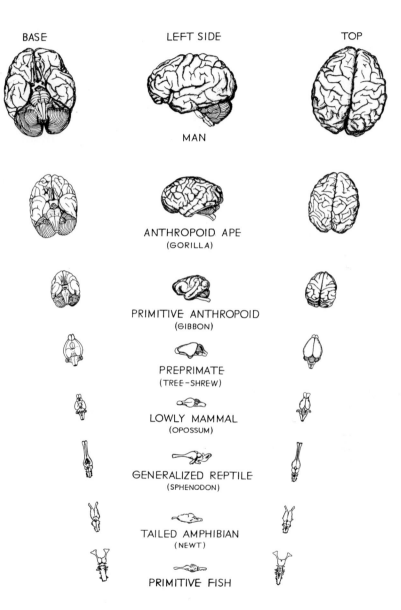

BASE   LEFT SIDE   TOP

MAN

ANTHROPOID APE
(GORILLA)

PRIMITIVE ANTHROPOID
(GIBBON)

PREPRIMATE
(TREE-SHREW)

LOWLY MAMMAL
(OPOSSUM)

GENERALIZED REPTILE
(SPHENODON)

TAILED AMPHIBIAN
(NEWT)

PRIMITIVE FISH

Fig. 4. THE RISE OF THE HUMAN BRAIN

over all other living creatures with which they came into contact.

Cro-Magnon Hunters Appear in Europe

The first to appear in Europe who were unquestionably of our own type (Homo sapiens) were a superb race of hunting men. They came into Europe during the latest ice age.

"The . . . entry of true man into Europe was led by the most magnificently developed folk the race has ever produced. . . . Some of the males stood six-feet-five inches high; the shape of the face suggests keen intelligence, and in size of brain they equalled or even surpasssed modern man."[31]

"The highly evolved characters of the skeleton in this (Cro-Magnon) race are in keeping with the extraordinarily great cranial capacity. . . . This race, observed Keith, was one of the finest the world has ever seen."[32]

". . . The Cro-Magnon, . . . seem very largely to have replaced the earlier Neanderthal cave-dwellers. The physical and mental qualities of this new race are the wonder and admiration of all students of the pre-historic period."[33]

The new race was erect and towered over the stocky Neanderthals. (Compare the skeletons of the two types in Figure 5.) The Cro-Magnons do not appear to have had at first any radically new weapons, such as bows and arrows, or dogs. They simply outdid all other cavemen by making weapons of greater keenness,[34] utilizing them more skillfully, organizing themselves more effectively and rearing more children than their predecessors did. They must have seemed like scourging gods to the Neanderthal people whom they gradually superseded.

------

31 Martin Stevers, *Mind Through the Ages* (Garden City: Doubleday, 1940), p. 91.

32 Henry Fairfield Osborn, *Men of the Old Stone Age* (New York: Scribner's, 1918). The observation of which he speaks appears in Sir Arthur Keith, *Ancient Types of Man* (New York: Harpers, 1918), p. 71.

33 C. C. Huntington, *The Geographic Basis of Society* (New York: Prentice-Hall, 1933), p. 542.

34 "The Solutreans (a Cro-Magnon culture) attained a high degree of skill. Their spearheads, lances and other flint products were the finest of the Stone Age." —Chesley Manly, *One Billion Years*, p. 32 of Chicago Natural History Museum reprint of his newspaper article.

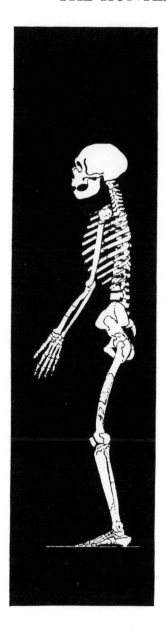
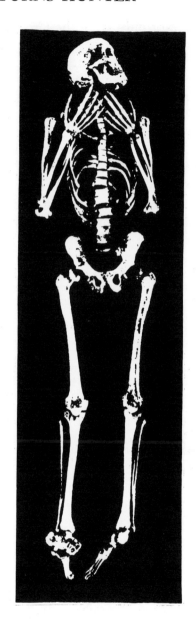

Fig. 5.   COMPARATIVE VIEW OF NEANDERTHAL AND
CRO-MAGNON SKELETONS
     View of a Neanderthal Skeleton (left) and of the skeleton of a very
tall member of the Cro-Magnon race (right). Both figures are approximately one-seventeenth life size. (*From Men of the Old Stone Age* by
H. F. Osborn, New York, Scribner's, 1918, p. 297.)

Not only were the Cro-Magnons impressive physical specimens, they more than had brains to match.[35]

"The intelligence and the artistic and spiritual qualities of the Cro-Magnon race are most surprising. With a body like our own and a brain at least as large as ours, superior individuals of this race would have been capable of becoming senior wranglers at any of our modern universities."[36]

"The Cro-Magnon . . . brain was much larger than the average of today. . . . It was . . . a superb race, both physically and mentally."[37]

These were the most intelligent human beings in more than a million years of selection for intelligence; the mightiest hunters in a million years of hunting men. The woolly rhinoceros and the

———

[35] "The Cro-Magnon . . . brain is much larger than the average of today. . . . It was . . . a superb race, both physically and mentally." — Herdman F. Cleland, *Our Prehistoric Ancestors* (Garden City: Doubleday, 1928), p. 23.

"The adult male buried at Cro-Magnon measured 5 feet 11 inches in life; five men at Grimaldi measured from 5 feet 10½ inches to 6 feet 4½ inches, averaging 6 feet 1½ inches. The tallest men now on earth, certain Scots and Negroes, average less than 5 feet 11 inches. . . . This race was not only tall, but clean-limbed, lithe and swift.

"Their brains were equally large. Those of the five male skulls from Grimaldi contain from over 1,700 to nearly 1,900 c.c.—an average of 1,800 c. c.; that of the old man of Cro-Magnon, nearly 1,600 c.c.; of a woman there, 1,550 c.c. If these individuals were not exceptional, (regarding this point Collie lists 105 Cro-Magnon skeletons, found in nine different sites, ranging from Wales to Czechoslovakia) the figures mean that the size and weight of the brain of the early Cro-Magnon people was some fifteen or twenty per cent greater than that of modern Europeans." —Alfred L. Kroeber, *Anthropology* (New York: Harcourt, 1958), pp. 27-28.

". . . to this number must be added several incompletely known skeletons from the Aurignacian deposits of North Africa." —George L. Collie, "The Aurignacians and Their Culture," *Beloit College Bulletin*, 1928, Vol. 26, No. 2, p. 27.

"All the skulls are remarkable for their great size, due partly to the great stature of the individuals.

"The skulls exhibit a certain degree of variation, but . . . the differences do not exceed the individual variations to be met with in relatively homogeneous groups." —Marcellin Boule and Henri Vallois, *Fossil Men* (New York: Dryden Press, 1957) translated by Michael Bullock, pp. 294-5.

[36] *Asia*, June, 1924, p. 431.

[37] Cleland, *Our Prehistoric Ancestors, op. cit.*, p. 23.

mammoth were fair game to them. So were any other sources of good, red meat: cave bear, bison, reindeer, wild horses.

". . . the Upper Paleolithic (Cro-Magnon) men hunted the same choice valleys used by Neanderthal Man. . . . But they were special and skillful hunters. At Solutre in France, they killed a hundred thousand horses, and at Predmost in Czechoslovakia, a thousand mammoths. The Neanderthals were surely able and valiant in the chase, but they left no such massive bone yards as this."[38]

The Cro-Magnon peoples were not only great hunters; they were superb artists.

"The existence of arrow straighteners implies the existence of the arrow itself and thus it would seem that Aurignacian (Cro-Magnon) man had already invented that powerful weapon—the bow. Armed with this he was able to take full advantage of the favourable circumstances by which he was surrounded. Life was easier and among its amenities may be counted a certain amount of leisure. Hence we now witness the birth of the fine arts. Sculpture and drawing almost simultaneously make their appearance, and the best examples attain to so high a pitch of excellence that enthusiastic discoverers have spoken of them as superior in some respects to the work of the Greeks."[39]

"No such skill in picturing living animals appeared again until the days of the ancient Greeks, perhaps 12,000 years later; and the Greeks had comfort and a wealth of materials for their work. These paleolithic artists achieved their effects with tools made of chipped flint, and they worked half-naked and shivering, with light from a lamp made by floating a moss wick upon melted fat in a hollowed out stone!"[40]

"We may pause to seek some reason for the momentous change when the Neanderthal appears to have bowed before the Cro-Magnon. The real secret in the failure of the old race and the success of the new may be found in the brain. It was the increased brain power of the Cro-Magnon which produced the supremacy of this great race. It was this power which gave

---

38 Howells, *Mankind in the Making, op. cit.*, p. 206.
39 William J. Sollas, *Ancient Hunters*, 3rd ed. (New York: Macmillan, 1924), p. 366.
40 Stevers, *Mind Through the Ages, op. cit.*, p. 97.

Europe its first pioneers in art and, for all mankind, opened the doors of creative imagination and appreciation of beauty in the world."[41]

"The Cro-Magnon people must have developed a high power of mental concentration to be able to observe and reproduce so closely."[42]

It is important to visualize the state of evolution attained by man toward the end of the latest great ice age. In Europe and in parts of Asia and Africa the prevailing humans were typically tall and very intelligent. Lacking agriculture, writing and other great cultural gains which man was to make in time, they remained prehistoric cavemen. But as men they have been seldom equalled, never surpassed. Here were the magnificent products of a million and more years of intense natural selection. Biologically the human race would rise no higher unless severe natural selection were to return or man were to employ an effective alternative to it.

In those days and in those areas few second-rate humans existed, for such were commonly eliminated by the rigors of nature and of competition. Survival was usually for a few of the fittest alone. Today, twelve thousand years later, with the ice age ended, agriculture providing food at least a thousand times more abundantly, and natural selection much enfeebled, the earth is overrun with masses of humans, mostly inferior to those. But in the time of the Cro-Magnon peoples there were no masses and no proletariat; of the Cro-Magnons there were relatively few but these few were superb.

Yet these once mighty people, who dominated Europe for at least 25,000 years—longer than the whole period of written history and the rise and fall of civilizations—finally failed. We know some of the reasons. The race deteriorated. Later specimens were not the equal of their forefathers. The climate grew milder and more humid;[43] large game grew scarcer and the Cro-

----

41 Frederick Tilney, *The Master of Destiny* (Garden City: Doubleday, 1930), p. 259.

42 Kroeber, *Anthropology, op. cit.*, p. 172.

43 ". . . geologists believe that the Wurmian glaciers began melting away some twenty-five thousand years ago. They were reduced to substantially their present extent of covering Greenland, Antarctica and the

Magnons had to resort more and more to fishing. Their descendants, or at any rate their successors, showed far less skill at weapons-making. Other races, with superior techniques, moved in. Between 10,000 and 7,000 years ago the Cro-Magnons were substantially replaced throughout their domain.

"The fate of the Cro-Magnon race was no exception to what had gone before or what would follow many times thereafter. Race after race . . . rose and became master, declined and passed into final extinction. As the day of Cro-Magnon ascendancy waned, a new race invaded western Europe. The Old Stone Age came to its end approximately ten thousand years ago with the advent of Neolithic (New Stone Age) man. He developed a great innovation in manufacturing his implements, making his instruments better and more useful by polishing the stone."[44]

"The principal change, however, was an economic one, namely the introduction of a rudimentary knowledge of agriculture with the corresponding use of a variety of plants and seeds.

". . . the New Stone Age witnessed a domestication of animals which corresponded closely with that of present times and included cattle, sheep, goats, pigs, horses and dogs."[45]

". . . possibly a thousand farmers could exist on the land which would only support a single hunter. Thus, from the very nature of their industry the Neolithic people could scarcely fail to grow strong numerically, and consequently capable of forcing their way into fertile regions in face of whatever resistance the hunters might oppose."[46]

higher mountain ranges about 6800 B.C., or 8700 years ago." —Martin Stevers, *Mind Through the Ages, op. cit.*, p. 74.
44 Tilney, *The Master of Destiny, op. cit.*, pp. 79-80.
45 Tilney, *The Brain From Ape to Man*, Vol. II, *op. cit.*, p. 766.
46 Sollas, *Ancient Hunters, op. cit.*, pp. 586-87.
I grew up in a Northern village established in what had been, until a few years before, an Indian reservation. Indians and whites went to the same school, played together and, like most boys, fought among themselves at times. Some of us discovered that Indian boys whom we could not ordinarily whip could be whipped toward the end of a long winter. We didn't realize they were weak from undernourishment. We just knew they couldn't run as fast or fight as hard then. We knew that more Indian than white schoolmates sickened and sometimes had to be buried, about the time the snow began to melt. I suppose we were witnesses to one of the advantages which, in lands where winter comes, children of those who produce and store food had over the children of those who only got what they could find and take.

With the decline of the Cro-Magnons, a new thing appears in the evolution of man. As later sapiens types replaced earlier ones, there was no further upward surge apparent in the inherent physical or mental abilities of the newly dominant peoples. The cultural advantages of the successor peoples were enormous (polished stone tools; then agriculture and domesticated animals) but the people themselves appear not to have been intrinsically superior. If anything, the evidence indicates some regression in brain size and other characteristics.[47]

This quite possibly represents the first time in human evolution that peoples, because of sweeping cultural advantages, managed to replace other peoples despite the absence of innate superiority over them. Certainly it is the first known instance in which people of so advanced a type were displaced over so great an area. As such it may represent the turning point in the biological evolution of man; the passing of the peak of physiologic fitness of our species. For the first time cultural evolution may have prevailed over biological evolution. At any rate, cultural gains continued to accumulate; biological gains no longer did.[48] As a consequence, Homo sapiens ceased to increase in quality. Instead, he began to increase enormously in quantity.[49]

————

47 "It is not until the Neolithic, some 10,000 years before Christ, that we see the establishment of peoples with a lighter and smaller brand of skull, and with them come the beginnings of the modern world." —William Howells, *Mankind So Far* (Garden City: Doubleday, 1944), p. 193.

"We have . . . clear evidence that the size of the skull depends on the size of the brain and not vice versa. The only exception to this rule is the condition of . . . Crouzon's disease." —Clemens E. Bendar, *Journal of the American Medical Association*, Vol. 201, No. 3, July 17, 1967, p. 134.

48 It is quite possible that the precise turning point in the evolution of humankind—the time when natural selection was weakened until deteriorative influences could predominate over ameliorative ones—occurred *within* the Cro-Magnon peoples and became even more manifest in their successors. See reference 49 for a discussion of this.

49 "It is said that Paleolithic man doubled his numbers every 30,000 years. Today the world population doubles every 30 or 40 years—roughly 1,000 times as fast." —Jno. R. Platt, "The Step to Man," *Science*, Vol. 149, August 6, 1965, p. 610.

The evidences of man's evolution, as known today, indicate a cessation of increase in brain following the climax of Aurignacian culture. Why did a trend which persisted for millions of years, and was so clear-cut and significant, come to an end then and possibly go into reverse? So pivotal a matter justifies our best conjecture.

We know that brain size and intelligence tended to increase under the severe natural selection which food-gathering and hunting imposed. We know that the increase apparently ceased with the advent of mixed agriculture. It is not difficult to see why this should have occurred, for food production permitted millions with lesser brains to survive who would not have qualified for survival under the more rigorous selection of the hunting stage. But why, if brains are so important, were the brainiest people replaced by others with slightly smaller brains? Why didn't those with the biggest brains devise agriculture?

It is probable that the best conditions for the development of agriculture were not the very best for human development. So many of the plants and animals best suited to domestication were indigenous to a certain area of Asia Minor that it is likely that agriculture was almost inescapable for the occupants of that region, after the climate of the ice age had moderated so that these plants and animals could flourish there.

"It happens that just in those regions of Hither Asia where ancestors of wheat and barley grew spontaneously, there lived wild sheep, goats, cattle, and pigs." —V. Gordon Childe, *What Happened in History* (Baltimore: Penguin, 1964), p. 56.

Agriculture is more of a discovery than an invention. The fortunate people of the favored area, if not quite the brainiest, were almost so. They were bright enough to take advantage of the situation which their unique environment presented to them.

Food production will support more than a thousand times as many people as hunting. It also provides more leisure for further development of other techniques. Thus the people to whom agriculture was first given developed such advantages in numbers and other decisive matters that they could overwhelm neighboring peoples. If they hadn't better brains, they had so many more good brains at work, and so much more time for them to work on new devisements, or devote to combat, (except during periods of plenty, a hunter who diverted much of his time to combat exposed his family to possible starvation) that they could outstrip others who individually might be as well or better endowed intellectually.

It appears that, as the climate of the latest ice age moderated, the Cro-Magnons of Europe had to turn increasingly from big game hunting to fishing. The ingredients for agriculture were not present where they lived.

As the rapidly increasing agriculturists poured out of their homeland they had the advantage both of vastly greater numbers and more effectual use of land area. If, in addition, the European hunters were in a decadent condition, as seems indicated, the whole impact was evidently a challenge to which, despite their larger brains, they did not successfully rise. The brainiest people of today also are threatened with extinction (see Part II of this book) and it is a grave question as to whether or not they will rise to the lethal challenge.

CHAPTER III

HUNTER TURNED FARMER

The occurrence which lifted man out of the hunting stage was the discovery of agriculture. It may well have been an especially observant woman—later to enter folklore as the goddess Ceres—who first noticed that the discarded seeds of fruits and cereals sprouted luxuriantly on the refuse heaps and before long multiplied themselves many times over. From some such observation human beings discovered, after the severe climate of the latest Ice Age had moderated, that the planting of seeds brings forth increase. Man learned to sow as well as to reap and consequently harvested far better than before.[1]

The crops of grain attracted grazing animals, including wild goats. Of these, many were killed by the planters of the fields but a few kids may have been kept as pets.[2] Some of the young goats grew to full size and produced offspring of their own. Here was a way to multiply the capture! Men learned to encourage nature in her abundance and the key to this abundance was the planting of seeds and the breeding of animals.[3] In the lands south

---

1 "At about the time when the last great glacier was finally melting away [in the North], men in the Near East made the first basic change in human economy. They began to plant grain, and they learned to raise and herd certain animals. This meant that they could store food in granaries and 'on the hoof' against the bad times of the year. This first really basic change in man's way of living has been called the 'food-producing revolution'. By the time it happened, a modern kind of climate was beginning. Men had already grown to look as they do now. Know-how in ways of living had developed and progressed slowly. . . . Once the basic change was made—once the food-producing revolution became effective—technology leaped ahead. . . ." —Braidwood, *Prehistoric Man, op. cit.*, p. 20.

2 "With a single exception—that of the dog—the earliest positive evidence of domestication includes the two forms of (wild) wheat, the barley, and the goat." —Braidwood, *Prehistoric Man, op. cit.*, p. 109.

3 "Concerning the animal aspect of the 'food-producing revolution,' present evidence indicates that domestication of goats and sheep

of the Caspian Sea man began about ei⸻ ⸻housand years
ago to gain his living by the tending o⸻ ⸻ domestic animals. He became the only vertebrate ⸻ ⸻abundance for
himself.

In discovering the arts of agricultu⸻ ⸻an acquired a new
degree of control over his environment. ⸻was no longer so dependent upon the fortunes of day-to-day hunting. He became,
more than any other creature, a partner in nature's productive
processes and, above all other living things, the arbiter of his own
fate.

When man discovered that planting fed him more reliably
and much more abundantly than hunting, he was inclined to
attach himself to the soil he worked and to feel proprietorship
toward it. Thus did real property, as distinguished from personal
possessions, come into being.[4] By its use man was helped to escape
from savagery.

---

occurred in a central core area in southwestern Asia in prehistoric times,
probably about the 7th millennium B. C., cattle being domesticated somewhat later, and pigs even later.

"Domestication of the food-producing animals probably occurred
in village-farming communities in the Hilly Flanks area of southwestern
Asia; thus, cereal agriculture and the settled village are considered to
antedate the domestication of all animals except the dog.

"Present archeological data indicates that relatively intensive and
successful agricultural and stock-breeding (mixed-farming) societies developed in the Zagros hills (West and South of Iran) and their grassy
forelands . . . prior to the appearance of the earliest societies of this type
elsewhere." —Charles A. Reed, "Animal Domestication," *Science*, Vol.
130, Dec., 1959, p. 1638.

4 "Whereas to the savage his quarry is of value only when it is
dead and has lost its capacity to increase, with the farmer his domesticated animals and plants are carefully tended and sheltered from potential
enemies. They are no longer victims so much as property. By breeding
animals in captivity and appropriating their natural increase, man in effect
draws dividends from the cycle of animal life. . . . Moreover, within the
limits of naturally available foodstuffs, he could enlarge his animal stock,
while at the same time consuming its by-products and a proportion of
its increase. Even more profitable was the cultivation of plants, since
not only was the annual increase very much larger, but the control was
more direct, and the possibility of enhancing this increase through
selective breeding was correspondingly greater. All in all the principal
economic characteristic of farmers as opposed to foodgatherers is that,
whereas the latter consume whatever foodstuffs they can wrest from
wild nature, the farmers discriminate between capital and income, consuming the increase of their fields and herds, but encroaching on their

There were, and still are, crude agriculturists who ravage the land they till and then move on to virgin soil. There are also nomads who follow their herds of animals. There are even, in parts of the world such as Australia, remnants of primitive, stone-age hunters who roam widely in search of their food, as all men did throughout the million or so years of man's early development. These notwithstanding, farming became the predominant way to support human life and this enabled men to establish permanent homes. Also, laborious though it was, farming gave them far more leisure for other pursuits. Now that man's exceptional mind no longer had to be occupied chiefly in obtaining the next meal, it was applied to many new fields of endeavor. Like Antaeus, man gained power from his contact with the earth. The new abundance also enabled greater numbers to gain a livelihood in a given area and with this the population began to swell beyond all previous increase. Soon, instead of a few single dwellings, many clusters of dwellings began to appear.

"The food producing state seems to appear in western Asia with really revolutionary suddenness. It is seen by the relative speed with which the traces of new crafts appear in the earliest village-farming community sites themselves, and the remarkable growth in human population we deduce from the increase in sites."[5]

Much of the countryside became widely and permanently, even densely, occupied. Clusters of dwellings grew into villages, then into towns and later into cities.[6] Man, already the most favored of nature's subjects, had now in truth inherited the Earth.

Civilizations

As man's settlements grew in size and complexity, there began in certain fertile and populous areas of the world a marked transition away from barbarian ways of living. There came into

---

stock and seed only in direst necessity." —Grahame Clark, *From Savagery to Civilization* (London: Cobbett, 1946), p. 71.

5 Braidwood, *Prehistoric Man, op. cit.*, p. 105.

6 "As far as is known, the world's first cities took shape around 3500 B. C. in . . . the valleys of the Tigris and Euphrates." —Gideon Sjoberg, "The Origin and Evolution of Cities," *Scientific American*, Vol. 213, No. 3, Sept., 1965, p. 56.

being the first of those advanced conditions of urban culture which are called civilizations.

The earliest of these great transitions, made possible by the enhanced state of material well-being, probably occurred in Sumer on the fertile banks of the Tigris and Euphrates rivers about five and a half thousand years ago.[7] Among the components of civilizations are cities, governments and large-scale coordinated projects, such as irrigation or war. The Sumerians were the first to live in cities, around which they constructed walls, partly to protect themselves against rising waters, partly to ward off enemy attacks.

In Sumer spoken words came to be represented by visual symbols. In this way the first written language on earth was developed.[8] Civilization, communication by means of writing and the recording of history all emerged together.

There followed one of the most fruitful eras in human history. It produced the wheel.[9] It also ushered in large-scale smelting of metals, thus terminating the Stone Age in Sumeria.

"Knowledge of how to reduce copper ore, so that the metal could be conducted in a molten state into moulds prepared to the shape of tools required for daily use, marked the greatest advance in man's equipment as a craftsman since he first learnt to

---

7 ". . . the next era is that of the appearance of urban civilization in Southern Mesopotamia, about 3500 B. C. . . ." —R. J. Braidwood, "Near Eastern Prehistory," *Science*, Vol. 127, No. 3312, June 20, 1958, p. 1429.

"Civilization appeared as a special intensification of cultural activity which effective food production made possible. . . ." —*Ibid.*, p. 1419.

8 "It was probably toward the end of the fourth millennium B. C., about five thousand years ago, that the Sumerians, as a result of their economic and administrative needs, came upon the idea of writing on clay. Their first attempts were crude and pictographic; they could be used only for the simplest administrative notations. But in the centuries that followed, the Sumerian scribes and teachers gradually so modified and molded their system of writing that it completely lost its pictographic character and became a highly conventionalized and purely phonetic system of writing. In the second half of the third millennium B. C., the Sumerian writing technique had become sufficiently plastic and flexible to express without difficulty the most complicated historical and literary compositions." —Samuel Noah Kramer, *History Begins at Sumer* (Garden City: Doubleday, 1959), p. xix.

9 Regarding the wheel, see E. A. Speiser, "Ancient Mesopotamia," *National Geographic*, Jan., 1951, p. 61.

work flint. The use of metal tools enormously enhanced his
mastery of stone, wood and a whole range of substances."[10]

In this same period were engendered arithmetic, astronomy,
the arch and the dome.

All around the Sumerians were peoples in less advanced
stages of culture. How high must have been the hopes of the
Sumerians who recognized that they had broken through a previously universal level of primitive existence and contemplated
the new way of life which they had brought into being!

As the Sumerian cities grew, systems of government were
elaborated within them. Conflicts arose between cities over water
and land rights, and battles occurred. These increased in size
and ferocity as the populations grew and governments were able
to marshal more fighters. Where once there were tribal skir-
mishes, there now were major conflicts. Out of political organization came the grimmest by-product of civilization: organized
warfare.[11]

"The will to war was in existence long before the dawn of
civilization; the latter, therefore, is not the cause of war, but it

---

10 Clark, *From Savagery to Civilization, op. cit.,* p. 95.

11 ". . . Every one of the factors contributing to the rise of civilization served to increase the probability of conflict, and war itself, tapping in
the instinct of self-preservation one of the deepest springs of life, called
forth to the full the latent energies of mankind, speeding up the rate
of technical development and intensifying the integration of society.
The closer settlements became, the more sharply were frontiers defined,
and the greater the material progress in the richer lands, the stronger
the inducement to break in from the poorer ones. Conversely, it was
characteristic of advanced societies to reach out beyond the frontiers
to more distant markets and sources of raw materials. This not only
involved persistent 'colonial wars' at the expense of barbarian or less
highly civilized neighbors, but also led to major wars between rival
imperialisms on the same level of cultural development. At the same time
technical progress constantly improved the material apparatus of slaughter
—metallurgy gave keener and tougher weapons and chariots increased
the mobility of the warrior—while an ever-growing population served to
fill the ranks.

"War has thus played a dual role in the evolution of civilization,
having been in part the consequence and in part the cause of cultural
advance. Throughout history, indeed, success in war has been the prize
of civilization. . . . The noble savage, the proud and vigorous barbarian,
the cultured citizen, all have bowed the neck to the lethal onslaught of
enemies superior in the means of taking life." —Clark, *From Savagery
to Civilization, op. cit.,* pp. 103-4.

did provide the conditions which make calamitous wars possible."[12]

There also developed, chiefly in cities, the adverse effects of artificial living which, like war, we have not yet learned to control. Among them was the fact that natural selection became so impeded that predominance of the most capable people largely gave way to predominance of the most prolific.[13] The result was then and, by and large, has been ever since, an increasing proportion of people of lesser ability. This was particularly true in the late and highly urbanized stages of civilizations. These two influences, external war and internal vitiation, the first recorded plentifully in history, the other so insidious as scarcely to have been appreciated before a hundred years ago, appear to have contributed decisively to the downfall of Sumerian civilization.

"Of the Sumerians nothing more is heard . . . the race had gone, exhausted by wars, sapped by decay, swamped by the more vigorous stock which had eaten of the tree of their knowledge."[14]

––––––––

12 Sir Arthur Keith, *Evolution and Ethics* (New York: G. P. Putnam, 1947), p. 126.

13 ". . . with savages, the average includes only the more capable individuals, who have been able to survive under extremely hard conditions of life." —Darwin, *The Descent of Man, op. cit.*, p. 437.

However, as civilizations developed, the growing urban centers attracted more and more of the able population. In cities the rearing of children was more difficult and costly, so that many couples who had the understanding and foresight to control their multiplication did so to such an extent that their kind diminished in proportion to the rest. At the same time most of those who were deficient in understanding or forethought went on having children much as before.

"The great peasant masses . . . multiplied exceedingly." —Childe, *What Happened in History, op. cit.*, p. 131.

"There is a tendency of the best men in the country to settle in the great cities, where marriages are less prolific and children are less likely to live . . . there is a steady check in an old civilization upon the fertility of the abler classes; the improvident and unambitious are those who chiefly keep up the breed. So the race gradually deteriorates, becoming in each successive generation less fitted for a high civilization, although it retains the external appearances of one, until the time comes when the whole political and social fabric caves in and a greater or lesser relapse to barbarism takes place, . . ." —Sir Francis Galton, *Hereditary Genius* (New York, Horizon Press, 1952), pp. 347-48.

14 Sir Charles Leonard Woolley, *The Sumerians* (Clarendon: Oxford, 1929), p. 182.

The same influences have also helped to bring about the downfall of every civilization great and small which came after, excepting our own so far.[15]

——————

15 Here are comments of several authorities regarding the decline and fall of civilizations:

"Of twenty-eight civilizations that we have identified, eighteen are dead and nine of the remaining ten—all, in fact, except our own—are shown to have already broken down. The nature of a breakdown can be summed up in three points: a failure of creative power in the creative minority . . . an answering withdrawal of allegiance . . . on the part of the majority; a consequent loss of social unity in the society as a whole." —Arnold J. Toynbee, *A Study of History*, D. C. Somervell's abridgement of Vols. VII-X (London and New York: Oxford, 1946).

"Among the problems presented by the social evolution of Man the most conspicuous is that of the decay and ruin of all civilizations previous to our own, in spite of their having had every reason to anticipate continued success and advancement." —Sir Ronald Fisher, *The Genetical Theory of Natural Selection* (New York: Dover Publications, 1958), p. 206.

"Such was the high biological level of the selected stocks which attained the plane of civilization. But, as time passed, the situation altered. The successful superiors who stood in the vanguard of progress were alike allured and constrained by a host of novel influences. Power, wealth, luxury, leisure, art, science, learning, government—these and many other matters increasingly complicated life. And, good or bad, temptations or responsibilities, they all had this in common: that they tended to divert human energy from racial ends to individual and social ends.

"Now this diverted energy flowed mainly from the superior strains in the population. Upon the successful superior, civilization laid both her highest gifts and her heaviest burdens. The effect upon the individual was, of course, striking. Powerfully stimulated, he put forth his inherited energies. Glowing with the fire of achievement, he advanced both himself and his civilization. But, in this very fire, he was apt to be racially consumed. Absorbed in personal and social matters, racial matters were neglected. Later marriage, fewer children, and celibacy combined to thin the ranks of the successful, diminish the number of superior strains, and thus gradually to impoverish the race.

"Meanwhile, as the numbers of the superior diminished, the numbers of the inferior increased. No longer ruthlessly weeded by natural selection, the inferior survived and multiplied.

"Such are the workings of that fatal tendency to biological regression which has blighted civilizations." —Theodore Stoddard, *The Revolt Against Civilization* (New York: Scribner's, 1922), pp. 18-19.

"In that long period of human history during which our evolving and expanding hominid ancestors lived in small and tightly knit groups competing for territorial and technological success, the social organization promoted selection for intelligent exploration of possibilities, devotion and cooperative altruism: the cultural and the genetic systems reinforced each other. It was only much later, with the growth of bigger social units . . . that the two became antagonistic; the sign of genetic transformation changed from positive to negative and definite genetic improvement and advance began to halt, and gave way to the possibility and later the probability of genetic regression and degeneration." —Sir Julian Huxley, "Eugenics in Evolutionary Perspective," *The Eugenics Review*, Oct., 1962, p. 133.

"A reasonably high level of intelligence in the average man, plus superior ability in numerous individuals, is essential for the maintenance of a high state of civilization." —Edward C. Colin, *Elements of Genetics*, 2nd ed. (New York: McGraw, 1946), p. 336.

"Each new generation consists of an increasing percentage of the less able. . . . Hence more and more people become 'problem makers' and fewer are 'problem solvers'." —Dr. S. S. Visher, quoted in Introduction to Elmer Pendell, *Sex vs. Civilization* (Los Angeles: Noontide Press, 1967), p. 218.

". . . in the normal course of events mankind in a civilization deteriorates by way of birth rate differences until eventually the creative individuals are too few and too inadequately creative to maintain the social structure." —Pendell, *Sex vs. Civilization, op. cit.*, p. 168.

The causes of the decay of civilizations are, like the causes of disease, better known to us than to earlier peoples. This knowledge can be a very useful thing if effectively used. But we must take better thought for the future than the Sumerians, or the Egyptians, or the Greeks, or the Romans or any of the other civilized peoples did at the height of their power.

CHAPTER IV

FARMER INTO FABRICATOR

In practically all agricultural societies, farmers and millers found that each produced more if one concentrated on the growing of grain while the other specialized in milling it. In the time saved from having to hand-grind his own grain, a farmer could produce enough additional grain to pay the miller to grind for him and still have more for himself than by the old method. Similarly a miller, equipped to grind for a number of farmers, could receive in pay more grain than if he took time from milling to raise grain for his own needs. Farmers and millers both benefited from their division of the labor. Since each could produce more through specialization than by trying to produce all his necessities by himself, each man's wealth increased. The advantages of specialization applied also to many other fields of activity and greatly increased the quality and variety of products available to the participants.

However, each gained only if there were exchanges between them. If the exchanges were more or less penalized, as by levies of robber barons, racketeers or tax collectors, they both to that extent were made the poorer. If exchanges were entirely prevented, each was worse off than if they had not specialized. The essence of the system was not only division of labor but exchanges of goods and services. Of the essence also was the diminished self-sufficiency of each man; the growing reliance of each upon the other.

Simple exchanges and trading were carried on for thousands of years. The Phoenicians, the Greeks and the Romans were among the successful trading people. In the Roman world a vast and complex exchange economy prevailed. However, for more than 600 years after the disintegration of the Roman Empire, the western peoples lived in small, relatively self-contained communities. Exchanges, other than local ones, were few during the stagnant centuries of the Dark Ages.

The Scientific and Industrial Revolution

After the long and dark age which followed the fall of Roman civilization, an exchange economy began to function again on a widening scale as our own Western civilization developed. This time, instead of being limited largely to products of fields, mines and handcraft, exchange became increasingly augmented by output from a new means of production—the factory.[1] The wider the area in which a commodity could be marketed, the more specialized and efficient could be the production of that commodity. The more specialized the production, the more it lent itself to mechanization, that is, to production in a factory.[2] Factories are among man's latest and greatest tools. As an almost worldwide exchange economy developed under Western civilization, some factories came to occupy whole square miles of territory.

As this new and complex technique of production spread, it brought to millions of humans a level of material well-being so far above any previously known that it came to be referred to as the Industrial Revolution.

———————

1 Manufactories of a sort existed as early as Greco-Roman times. However, the essential difference in the factories which first made the fruits of the scientific revolution available to so many consisted of machines operated by steam power. Devised and controlled by human brains, power-driven machines increased the productivity of human hands manyfold.

2 "Passing to manufactures, we find here the all-prominent fact to be the substitution of the factory for the domestic system, the consequence of the mechanical discoveries of the time . . . in 1769—the year in which Napoleon and Wellington were born—James Watt took out his patent for the steam engine. In 1785 Boulton and Watt made an engine for a cotton-mill at Papplewick in Notts. . . . These two facts taken together mark the introduction of the factory system." —Arnold Toynbee, *Lectures on the Industrial Revolution in England* (Boston: Beacon Press, 1956), p. 63.

". . . men of our culture began to experience enough significant change in their daily lives to realize that they were entering a new epoch in human affairs. The realization came first of all to the people of England and Scotland, for they were the first large western communities to augment their wealth by losing their self-sufficiency." —Walter Lippmann, *The Good Society* (New York: Grosset and Dunlap, 1943), p. 162.

"A small handful of Scots and Englishmen—fewer than would be required for a football match—succeeded by their ingenuity in transforming the economic life of the country." —H. A. L. Fisher, quoted

Part and parcel of the revolution is the alliance of industry with science[3] and invention, which makes possible power-driven machines and apparatus in prodigious variety. The accumulation of earnings which are risked—so long as there is a reasonable chance of gain—makes possible even more and better machines and techniques. Through this progression the products offered often come to be so readily afforded and in such demand that their mass production becomes feasible. The economies of mass production then enable even greater numbers of people to afford the products.

"Two centuries ago not one person in a thousand wore stockings. . . . Now not one in a thousand is without them."[4]

"It is no exaggeration to say that the transition from the relative self-sufficiency of individuals in local communities to their interdependence in a worldwide economy is the most revolutionary experience in recorded history.

"The period from, say, 1776 to 1870, was the golden age of free trade and of political emancipation throughout the western world. It was an age when the reforming passion of men was centered upon the abolition of privileges, the removal of restraints,

---

in Paul Bloomfield's *Uncommon People* (London: Hamilton, 1955), p. 61.

"For more than a hundred and fifty years the revolution which converted relatively independent and self-sufficing local communities into specialized members of a great economy has been proceeding at an accelerating tempo. In the struggle for survival the less productive economy of self-sufficiency has not been able to withstand the superior effectiveness of a mode of production which specializes in labor and natural resources and thereby promotes the use of machinery and mechanical power. In some degree the worldwide division of labor has been checked by tariffs, and other barriers to the movement of capital and labor. But they have only retarded the process. Inside the nations which consider themselves most civilized there are now few communities left which are in any substantial sense self-sufficing. The self-sufficing household has virtually disappeared. Some nations, taken as a whole depend less on foreign trade than others, but none could even begin to maintain its present standard of life if it were isolated from the rest of the world." —Lippmann, *The Good Society, op. cit.*, p. 164.

3 ". . . the Age of Science dawned about the year 1700. . . . Galileo and Newton ended the era of speculative philosophizing and opened the age of scientific discovery." —Lee A. DuBridge, *Education in the Age of Science* (New Haven: Yale University, 1958).

4 *Encyclopedia Britannica*, 14th ed., "Industrial Revolution," p. 305.

the restricting of the authority of the state. It was an age when men were dominated by the conviction that it was by the method of emancipation, rather than by authoritarian planning and regulation, that mankind could most surely achieve its promise."[5]

Today, as the scientific and industrial revolution still continues, man has come to dominate the air, the world of plants and animals and, increasingly, the world of insect and bacterial life. He has learned to traverse vast oceans with ease and safety and to probe their uttermost depths. He is able to fly more swiftly than sound; to unleash the energies of fossil and nuclear fuels; to communicate with his fellows over thousands of miles with the speed of light. What emperor of old could travel at a mile a minute in a cushioned land-yacht? What ruler could fly at hundreds of miles per hour? What king could protect his children from dreaded smallpox?[6] These previously impossible privileges are now the common prerogative of the majority of people in the industrialized capitalist countries. Hundreds of millions enjoy comfort, luxury and freedom from want and numbing toil beyond anything previously experienced by man.

----

5 Lippmann, *The Good Society, op. cit.,* pp. 132 and 194.

6 "Bad as our urban conditions often are, there is not a slum in the country which has a third of the infantile death-rate of a royal family in the middle ages." —J. B. S. Haldane, "Science and the Future," *Daedalus,* 1924, p. 54. (A paper read to the Heretics, Cambridge, Feb. 4, 1923.)

CHAPTER V

THE INCREASE OF MAN

One of the fundamental facts of the world today is that mankind is increasing prodigiously in numbers. All his larger enemies are practically extinct. One after the other his epidemic diseases are being subdued. His food-producing abilities are multiplying enormously. Because of these and other accomplishments, man has been able to double and then re-double his number upon the earth, over and over again. Each twenty-four hours at least 172,000[1] are added to the earth's population. This is equivalent to creating another city larger than Geneva every morning. In the First World War something like ten million were killed in the course of four years. Today this many additional lives come into being in about two months. In the past twelve months sixty-three million more—a number greater than the population of the United Kingdom or West Germany— came to occupy the globe with us.[2] This year there will be sixty-five million more—more than a million per week; another India every eight years.[3] The first billion in population took more than 50,000 years of increase of Homo sapiens. The second took about 100 years. The third took less than forty years.[4] The world is filling up with people and the rate of increase is itself increasing.[5] One of every twenty men who ever lived is alive today.[6] Man has become an explosively successful biological phenomenon.[7]

---

1 *U.S. News*, Vol. 57, Dec. 19, 1964, p. 393.

2 *Ibid.*

3 Susan Michelmore, *Sexual Reproduction*, (New York: Natural History Press, 1964), p. 208.

4 G. W. Beadle, "The Uniqueness of Man," *Science*, 125:9-11, Jan. 4, 1957.

5 "If today's growth rate persists, the present population of about 3.3 billion will double in the next 33 years." —*Wall Street Journal, Pacific Coast Edition.* Vol. SXXXIX, No. 54, Sept. 16, 1968, p. 1.

6 Harrison, *What Man May Be, op. cit.*, p. 109.

7 "The great masses of men are just now arriving. And they really

The accompanying graph (Fig. 6) shows the upsurge in human population made possible by the scientific-industrial revolution, beginning after 1700 A.D.

An earlier upsurge began as a result of the food-producing revolution.[8] Population increased from the 1 to 2 millions at the beginning of agriculture (8000 B. C. approximately) to 200 to 300 millions at the beginning of the Christian era. However, the increase occurred over thousands of years instead of hundreds and, while proportionately very great at the time, was numerically so small that it is hardly visible on a graph of the scale of this one.

---

are arriving—the net increase in humanity during the nine-month gestation period which ended today would populate the cities of New York, Chicago and Philadelphia, plus Los Angeles, Detroit and Baltimore, plus Cleveland, St. Louis and Washington, plus Boston, San Francisco and Pittsburgh, plus Milwaukee, Houston and Buffalo, plus at least any other American city you choose. Such a mob is large enough, literally, to join hands around the earth. It is also a notably hungry, brutal and powerful mob—and it is likely to become more so as its numbers increase." —*Saturday Evening Post*, Vol. 223, Aug. 6, 1960, p. 33. J. D. Williams, Head of Mathematics Division, RAND Corporation.

"The yearly increase in the earth's population is today greater than the total population of the world at the dawn of agriculture some 8000 years ago." —Jas. Bonner, *The Bulletin*, California Institute of Technology, Jan. 21, 1965, p. 10.

8 Philip M. Hauser, "Demographic Dimensions . . .," *Science*, Vol. 131, No. 3414, June 3, 1960, p. 1614.

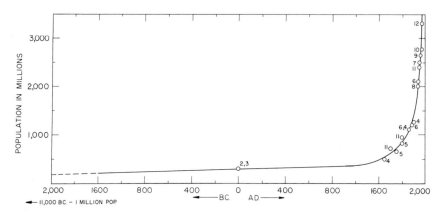

INCREASE OF WORLD POPULATION

GRAPH REFERENCE NUMBER	YEAR	POPULATION IN MILLIONS	SOURCE
1	1 million B.C.	⅛ (125,000)	*Scientific American,* Vol. 207, Sept. 1962, p. 268.
2	11,000 B.C. 1 A.D.	1 300	J. M. Luck, "Man Against His Environment," *Science,* Vol. 126, November 1, 1957, p. 126.
3	1 B.C.	300	Julian Huxley, *World Population.*
4	1650 1850 1900 1950	500 1,100 1,650 2,500	Berrill, *Man's Emerging Mind, op. cit.,* p. 208.
5	1750 1800	660 836	Alexander M. Carr-Saunders, Fig 7, *World Population; Past Growth and Present Trends,* 1936, Oxford.
6	1857 1947 1957	1,200 2,100 2,400	Frederick Osborn, *Science,* Vol. 125, March 22, 1957, pp. 531-34.
7	1860 1950	1,250 2,500	U.N. Dept. of Economics and Social Affairs, United Nations, N.Y. *U.S. News,* Vol. 45, August 29, 1958, p. 48.
8	1933	2,000	*Harvard Business Review,* July, 1957, p. 19.
9	1954	2,652	*U.N. Demographic Yearbook,* United Nations, N. Y., 1963.
10	1957	2,700	*Science News Letter,* Vol. 73, June 1, 1957 (U.N. Figure), p. 343.
11	1650 1700 1800 1950	545 728 906 2,400	E. S. Deevy, *Scientific American,* 203:195, 1960.
12	1963	3,300	*U.N. Demographic Yearbook, op. cit.*

CHAPTER VI

THE INCREASE OF MAN'S BRAIN IS ARRESTED

"Civilization has not increased the mental capacity of its subjects; what it has done is to supply men with the opportunities, the leisure, and the means to develop the mental gifts already attained by man while living in a state of nature."[1]

If we take our eyes from the fascinating contemplation of the knowledge and techniques which enable man to increase so mightily, and examine more closely the character of the increase itself, we perceive certain processes in operation which need to be recognized and dealt with if the future of man is to equal the present.

With each easement of man's once precarious situation, the severity of natural selection diminished. At first only a few more individuals, slightly less able than the best, could manage to survive and reproduce along with the ablest. However, as time and easier circumstances brought further security to man, more and more of the less able could and did survive and multiply.

". . . in Neanderthals 40% died as children of eleven years or less; but in later (Old Stone Age) man, only 24.5% died this young."[2]

". . . today in European countries 95% of the children born alive survive to reach the age of reproduction. The deaths of the 5% who do not survive are undoubtedly highly selective against physical abnormalities and weaknesses, but they are certainly not selective for the higher qualities. . . . Today the processes of selection affecting man's higher qualities operate not through

---

1 Sir Arthur Keith, *Evolution and Ethics*, op. cit., p. 88.
2 Weston La Barre, *The Human Animal*, op. cit., p. 101, (quoting from Franz Weidenreich, "The Duration of Life in Fossil Man . . .," *Chinese Medical Journal*, LV, Washington, D. C., 1939; pp. 34-44; also in *Anthropological Papers*, pp. 194-204).

73

deaths, but through differentials in the number of children born to people of different genetic types."[3]

Thus the effect of natural selection became so weakened that, among other things, it no longer resulted in further increases in the human brain. Now and then an increase of brain would occur, as in the past, but these increases no longer had the same selective advantage. As a consequence, that great and powerful organ in which, more than in any other way, humans excelled all other forms of life, ceased to develop further.[4]

Since the time of the hunters and the end of the ice age, one of the greatest helps to the survival of those with lesser brains was the development of farming. It was not that those with full mental equipment did not continue to farm. They did better than the others as a rule. It is just that farming, better than hunting, enables parents to protect and sustain children, even those with poorer brains, until they can reproduce. In a state of nature, many of these would have been picked off early by predators or enemies, or have been the victims of infanticide or starvation in

---

3 Frederick Osborn, "Galton and Mid-Century Eugenics," *Eugenics Review*, April, 1956, p. 16.

4 "We know that in the late Villafranchian era, when Australopithecus lived, hominid brain size was still around 650 cc., while some three or four hundred thousand years later, brain size had increased to 1500 cc. This was an extraordinarily rapid rate of evolution. But then the curve suddenly flattened out, and in the last 50,000 to 100,000 years there has been no increase in brain size at all." —Ernst Mayr, "Comments," *Daedalus*, Summer, 1961, p. 467.

"As far as is known, no structural changes occurred in the human brain during the ten to fourteen centuries constituting the barbaric interval, (primitive agriculture, domestication of animals, weaving of rough cloth) except possibly a slight retrogression in size. From the scientific evidence available, man's encephalon reached its full physical stature somewhere in the late Paleolithic (stone-age) period. . . . Very much the same is true from the functional aspect. The men living 10,000 to 15,000 years ago possessed a fairly high order of intelligence as well as physical ability." —A. M. Lassek, *The Human Brain* (Springfield, Ill.: Thomas, 1959), pp. 162-63.

"All the members of a community benefit from the technological and other achievements of the superior individuals and this helps the below-average individual, provided he is not too far below average, to make a living and to reproduce as successfully as the above-average one. . . . All I want to point out is . . . the almost abrupt flattening out of an exceedingly steep evolutionary advance. . . ." —Mayr, *Animal Species and Evolution, op. cit.*, pp. 652-53.

times of great scarcity. In farming very simple folk could make shift and increase their kind. They might not even be adept at adding or subtracting but they could multiply as never before.

On farms, more than in hunting, children could help to support themselves, so that early in life they became no particular burden. No previous pursuit had been so conducive to human increase as the practice of agriculture. And of course those who would have been eliminated under rigorous natural selection showed the greatest proportionate increase.

Yet even the agricultural revolution was subsequently outdone, as an incubator and supporter of the less fit, by the scientific and industrial revolution. Modern mass-production techniques permit a person of poor mind to do one thing over and over until he or she is extraordinarily adept at it. In fact, in this circumstance a limited mind can be an asset, enabling one to remain at a repetitive task until a facility is gained which one with a more ranging mind would not be likely to acquire. This was the reason a genius such as Henry Ford could, by breaking down complex procedures into their simplest components, and then assigning only one simple operation to a laborer, take many dull and almost unskilled individuals, shorten their work-day, raise their pay to levels never before known and still enable them to produce a profit.

There are other ways in which the industrial revolution may surpass agriculture as a means of sustaining the less able. One of these arises from the fact that a substantial part of its great wealth-producing powers is channelled, by charities and governmental relief, into the preservation and multiplication of those unable or unwilling to maintain themselves even by performing simple tasks.

Accordingly, as man changed his ways of gaining a living, there developed within his settlements increasingly effective social and economic shields. This not only permitted him to gain in numbers, it also served to shelter retrograde variants who would have been unable to survive long in the isolated families of the hunting stage. Today, in civilized countries, so complete is the shielding and so reduced are man's enemies that millions grow up and breed in spite of weaknesses and deficiencies which would have caused their early death during more stringent times.[5]

"With savages, the weak in body or mind are soon eliminated; and those that survive commonly exhibit a vigorous state of health. We civilized men, on the other hand, do our utmost to check the process of elimination; we build asylums for the imbecile. . . ; we institute poor-laws; and our medical men exert their utmost skill to save the life of every one to the last moment. . . . Thus the weak members of civilized societies propagate their kind. No one who has attended to the breeding of domestic animals will doubt that this must be highly injurious to the race of man. It is surprising how soon a want of care, or care wrongly directed, leads to the degeneration of a domestic race; but excepting in the case of man himself, hardly any one is so ignorant as to allow his worst animals to breed."[6]

"*We probably save for reproduction*, by means of our advanced medical, industrial and social techniques, *more than one half of the people who in past times would have had their lines of descent extinguished* as a result of their genetic shortcomings."[7]

In man, nature produced not the largest or swiftest or strongest creature but the one with a large and most effective brain.[8] So effective is his brain that man, instead of continuing to be dominated and gradually improved by the environment, has devised ways to dominate and alter the environment itself. The more he modified his environment, the less the environment could modify him. With this turning of the tables, the whole-

---

5 "Today the vast majority of persons born, even when subnormal, are enabled to reach maturity. . . . Moreover, under modern conditions, foresight, conscience, and competence seem to express themselves rather by successful restraint in reproduction than by abundance of offspring. Thus the earlier trend of natural selection is to some extent reversed. So long as such conditions prevail human populations must become ever more defective in their genetic constitution. . . ." —Herman J. Muller, *Human Genetic Betterment*, an address before American Humanist Assn., Mar. 29, 1963.

6 Darwin, *The Descent of Man, op. cit.*, p. 501.

7 H. J. Muller, "Life," *Science, op. cit.*, pp. 1-9.

8 The brain of man is actually larger than that of any other animal, however large, except the elephants and whales. Even compared with these, the effectivity of man's brain and its proportion to his body size is greater.

some influence of natural selection became less and less effective upon him. Nature, which had begun with a tiny blob of living protoplasm and, working through selection, shaped and reshaped it through countless generations until it became the wondrously intricate and successful creature called man, was now largely prevented from improving him further.

For the last several thousand years, man has at best stood still in his own physical and mental development, diverted by the cumulative cultural advances which a relatively few intelligent individuals have given him. Man of today, the creature who can soar to the moon, the manipulator of the powers within the very atoms themselves, exhibits no better brain or body than his late Stone Age ancestors who lived in caves thousands of years ago.[9] He has lost his main evolutionary drive. He has substantially freed himself from natural selection without employing an effective alternative to it. As a consequence he has become, at this stage in his existence, an evolutionary derelict. The direction of his drift is, if anything, downward.[10]

----

9 "There is no scientific evidence available that any structural or physiological changes have occurred basically in the brain of man since civilization began. Growth of this organ reached its peak and ceased in the late stage of savagery, perhaps 25,000 or more years ago." —Lassek, *The Human Brain, op. cit.,* p. 178.

The Cro-Magnon brain averaged 15 to 20% greater than that of modern Europeans." —Kroeber, *Anthropology, op. cit.,* pp. 27-8.

"In civilized nations the average capacity of the skull must be lowered by the preservation of a considerable number of individuals, weak in mind and body, who would have been promptly eliminated in the savage state. . . . This explains the fact that the mean capacity of the skulls of the ancient troglodites (cave dwellers) of Lozere is greater than that of modern Frenchmen." —Darwin, *The Descent of Man, op. cit.,* p. 437.

"So far as concerns the brain capacity of the skull, there is no evidence of increase. From the limited data at our disposal, we must infer that the people who occupied Western Europe at the close of the ice-age stood distinctly above their successors of today in the matter of brain size." —Sir Arthur Keith, *Concerning Man's Origin* (London and New York: Putnam's Sons, 1928), p. 160.

10 There are several schools of opinion regarding this matter of retrogression. One school holds that it was only within the past 100 years or so that adverse influences began to preponderate over favorable ones. This is probably true within our own Western civilization, taken by itself.

Others hold that the physiological decay of man began with the establishment of cities and civilizations. Some of these, more explicit,

"Though undoubtedly man's genetic nature changed a great deal during the long proto-human stage, there is no evidence that it has been in any important way improved since the time of the Aurignacian (Cro-Magnon) cave man. What has been improved since then is the tools of action and thought and the ways of accumulating and utilizing experience, and these improvements have had truly prodigious results in a very brief period of time. Indeed, during this period it is probable that man's genetic nature has degenerated and is still doing so."[11]

---

feel that adverse influences came to predominate during the later, highly urbanized stages of civilizations. This seems valid if we limit our view to the record of written history.

In this book the position is taken that the physiological decline of man probably began with the end of the latest ice age and that the most demonstrable decline began with the entrance of agriculturists into Europe. This is based on the findings that the average physique and relative brain size of the early and middle Cro-Magnon hunters were superior to the average known anywhere, at any time, before or since.

It is immaterial to an understanding of man whether the peak of mental fitness occurred with the Cro-Magnons or with, say, the classical Greeks (whose brain sizes are not determined). The fact is that we have not on the average improved intrinsically since distant times, and there has probably been some retrogression.

Yet we could probably find, among the millions with European ancestry spread across the earth today, more specimens who would equal in every respect the best of the Cro-Magnons than the Cro-Magnons ever could have marshalled at any one time. This fact gives man a tremendous potential. However, the presence of a few fine specimens in the herd should not blind us to the fact that the herd itself, once small and largely purebred, is now huge and made up mostly of scrubs.

11 Sir Julian Huxley, *Evolution in Action* (New York: Harper, 1953).

We can hope that the shrinkage in brain volume is compensated by increased efficiency of the remaining brain mass, but I can find no evidence to support the hope.

"Sheer size in brains is important for two reasons. Most obviously, a small brain simply cannot hold as many brain cells as a large one. Less obvious, but more important, is that the true quality of a brain must be measured by the complexity of the linkages between cells. Inasmuch as the possible number of linkages goes up very rapidly as the brain gets larger, it is clear that a big brain can be a much more sophisticated instrument than a small one." —F. Clark Howell, *Early Man* (New York: Time-Life Books, 1968), p. 82.

Or, one might hope that the shrinkage has occurred primarily in the older, more primitive parts of the brain. Actually, it is these parts which are more persistent. Apparently the shrinkage, where it has occurred, is chiefly in the more recently acquired, higher brain area (the neocortex).

"The new world of expanding mind develops in new expanses of cortex, the neocortex. These, . . . enlarging with evolution, . . . might . . . be called "thinking areas; for, as they increase from lower vertebrates through man, so do the powers of choice, experiment, reasoning, imagination, and . . . thought. . . . The size of these thinking areas in man is what makes him anatomically human." —Elliott, *The Shape of Intelligence, op. cit.*

The relation between size of brain-case and size of brain is absolute in each normal individual. This does not mean perfect relationship between brain size and intelligence. Nevertheless, if allowance is made for body size and slight differences in proportion between cortex and other parts of the brain, cortical size and indeed total brain size and intelligence are statistically positively correlated, as would be expected since the brain is the central instrument of intelligence.

"Averages calculated from groups, e.g., scholarship and prizemen, average, and below average ability, show statistically that there is a small though measurable correlation between size of head and a high intelligence. The relation becomes still more apparent when the heads of the intellectual classes are compared with those of the lower classes, and more particularly the inmates of workhouse infirmaries and congenital idiots." —*Encyclopedia Britannica*, 1961 ed., "Brain-I. Anatomy of the Brain. Subtitle: Weight of the Brain," p. 14.

"Eleven studies have been made of the relationship between intelligence and . . . cranial capacity. In all instances, the correlations have been positive, . . . ranging from .08 to .34." —Leona E. Tyler, *The Psychology of Human Differences*, 2nd ed. (New York: Appleton- Century-Crofts, 1956), p. 422.

A number of people, by citing rare individual exceptions to the rule, becloud—at least in their own minds—the fact that this positive statistical correlation between brain size and intelligence exists. Ascertainable regression in *average* brain size has occurred, at least among those of European descent. While not too much should be made of it, a shrinkage in average brain size, occurring among the largest-brained groups, can hardly be regarded as evidence of mental progress in man. Being contrary to the principal direction of primate evolution, it must be taken as strong evidence of retrogression.

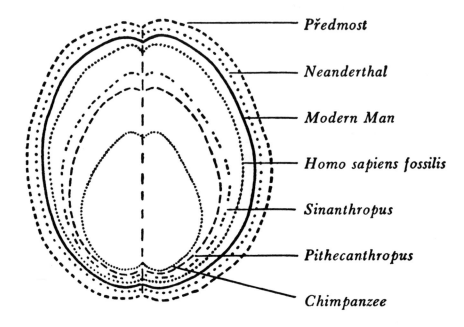

Fig. 7   A COMPARISON OF BRAIN SIZES

Chimpanzee, 400 c.c. Pithecanthropus, *860 c.c.* Sinanthropus, *1075 c.c.*
Homo sapiens fossilis, *1300 c.c. Modern man, 1400 c.c. Neanderthal man,
1450 c.c. Predmost, 1500 c.c. (a Cro-Magnon cognate).*

(From *Introduction to Physical Anthropology* by Ashley Montagu,
Springfield, Illinois, Thomas, 1945. Also in Montagu's *Man: His First
Million Years,* Cleveland, World, 1957.)

Chapter VII

PROBABLE CAUSES OF THE CESSATION OF BRAIN INCREASE

1. The Increased Proportion of Those With Lesser Brains

Natural selection is an elegant and euphemistic term for one of the cruelest schemes in all of nature. It requires excessive reproduction and follows it with selective destruction among the disadvantaged young. When operating at highest intensity, as it did upon man for the million or more years when it was lifting him above all other creatures, natural selection permitted only a fortunate few of the best adapted in each generation to survive until they could reproduce. Repeatedly it condemned to early death the great majority of all who knew life. Natural selection was both crude and cruel; yet for man it was especially constructive. Every comprehending human of today is the descendant and beneficiary of those thousands of generations in which the most capable and intelligent individuals were the ones who managed best to survive and hand on their capability and intelligence. This is the basic reason the intelligence of even the mediocre today is so far above that of all other living things. Yet man, more than any other creature, has now escaped from this cruel but salutary influence.[1]

In proportion as the influence of natural selection was enfeebled, a sinister counter-influence was enabled to assert itself. This is the grim fact of life that harmful mutations occur far

_____

1 ". . . in the United States and among other peoples of western European descent, among whom the death rate has gone down to a point where more than 90 percent of us survive beyond our thirtieth year, death has ceased to be the major factor in determining which types shall survive and have children. Natural selection by deaths has come almost to a halt." —Frederick H. Osborn, *Preface to Eugenics* (New York: Harper, 1951), p. 237.

more frequently than beneficial ones.[2] In fact, more than 99% of all mutations are harmful.

In nature, changes in the inherent character of creatures occur blindly[3] and most of the changes harm their possessors to some extent. Natural selection, when not interfered with, tends to weed out these harmful mistakes. If they are not weeded out, they multiply in the succeeding generations. When this occurs the result, over the years, is a cumulative increase of inferior qualities and a degeneration of the type.[4]

———

2 "Most mutations are more or less injurious to the organism. It is not hard to see why this is so. The injurious character of most mutants is not an attestation of inherent perverseness in nature. A change in a gene is expected to be injurious for the same reason for which random rearrangements of wires in a radio set would spoil it more often than improve it." —Theodosius Dobzhansky, "Genetic Loads in Natural Populations," *Science*, Vol. 126, August 2, 1957, p. 191.

3 "Genes are giant molecules, and their mutations are the result of slight alterations in their structure. Some of these alterations are truly chance rearrangements, as uncaused or at least as unpredictable as the jumping of an electron from one orbit to another inside an atom; others are the result of the impact of some external agency, like X-rays, or ultra-violet radiations, or mustard gas. But in all cases they are random in relation to evolution." —Sir Julian Huxley, *Evolution in Action* (London: Gollancz, 1953), p. 38. Based on the Patten Foundation lectures at Indiana University, 1951.

4 "In all organisms so far investigated, deleterious mutations far outnumber useful ones. There is an inherent tendency for the hereditary constitution to degrade itself. That man shares this tendency we can be sure, not only from analogy but on the all-too-obvious evidence provided by the high incidence in 'civilized' populations of defects, both mental and physical, of genetic origin.

"In wild animals and plants, this tendency is either reversed or at least held in check by the operation of natural selection. . . . In domestic animals and plants, the same result is achieved by our artificial selection. But in civilized human communities of our present type, the elimination of defect by natural selection is largely (though of course by no means wholly) rendered inoperative by medicine, charity, and the social services. . . . The net result is that many deleterious mutations can and do survive, and the tendency to degradation of the germ-plasm can manifest itself.

"Today, thanks to the last fifteen years' work in pure science, we can be sure of this alarming fact, whereas previously it was only a vague surmise. Humanity will gradually destroy itself from within, will decay in its very core and essence, if this slow but relentless process is not checked." —Sir Julian Huxley, *Man in the Modern World* (London: Chatto, 1947), p. 60.

"Those insufficiently familiar with genetic principles might suppose

Few who are aware of the facts maintain that natural selection results any longer in survival preponderantly of the most intelligent humans. The earth swarms with the evidences to the contrary.

"The intelligence and character of the masses are incomparably lower than the intelligence and character of the few who produce something useful for the community. . . ."[5]

As the enfeeblement of natural selection proceeded, the influence of the preponderance of harmful mutations—including those in the direction of reduced brain capacity—grew at an accelerating rate, feeding upon itself.[6] After the establishment of cities

---

that, in view of the relaxation of natural selection . . ., the human species would now stop in its tracks in regard to things biological. . . .

"But they fail to reckon with mutation. This goes on anyway, . . . the genetic constitution of a species does not maintain its status quo when selection is artificially relaxed or removed; it drifts inevitably backwards. This is because, as we have seen, mutations are in great majority detrimental, in fact disorganizing. And so, as they continue to arise generation after generation they will, if not removed by the death or relative infertility of the individuals inheriting them, gradually accumulate in the population, weakening it more and more in all its characteristics of form and function, including its mental functions." —Hermann J. Muller, *The Future Physical Development of Man*, from lecture at Pennsylvania State University, April, 1960.

5 Albert Einstein, *Message in the Time Capsule*, 1939. See his *Ideas and Opinions* (New York: Crown, 1954), p. 18.

6 "Without effective selection eventual deterioration of any adapted species is inevitable by reason of the buildup of nonadaptive mutant genes." —Robert C. Cook, Editorial, *Eugenics Quarterly*, September, 1965, p. 130.

"By preserving the lives of some biologically unfit offspring, and by granting the privilege of reproduction to some heavily defective adults mankind is currently accomplishing a degradation of its stock of genes." —Oscar Riddle, *The Unleashing of Evolutionary Thought* (New York: Vantage, 1954), p. 84.

"In a state of nature, this problem of the accumulation of defective genes was often taken care of effectively by 'natural selection': the strong man, the united tribe, more frequently won out, and the defective perished or failed to reproduce as abundantly. Natural selection of this primeval variety is today completely incompatible with happiness, sympathy, or civilization itself. The only adequate alternative to it, however is some kind of thoroughgoing eugenics. . . . Failing this, a biological deterioration of the human race seems to be inevitable, and this might even endanger the dominance of man as a species." —Hermann J. Muller, *Out of the Night* (New York: Vanguard, 1935), p. 44.

"In almost all the industrially and socially advanced countries, the

and the further protection from the hazards of nature which they provided, accumulations of those who were much less capable began to appear in these centers.[7] With their greatly increased rate of survival and relatively uninhibited rate of reproduction, these have come to multiply their kind beyond all others. At the same time the growing pools of such people, rural or urban, have been further augmented by the retrograde mutations and unfortunate genetic combinations which occasionally occur in the

---

level of innate intelligence, and probably of other desirable genetic qualities, is decreasing generation by generation. Furthermore, we can be certain on theoretical grounds that the relaxation of natural selection brought about through our medical knowledge and social care must be causing a slow degeneration of the stock, through the accumulation of harmful mutations." —Sir Julian Huxley, in Introduction to *Human Fertility* by Robert C. Cook (New York: Sloane, 1951).

"Another way in which lack of a very highly developed social impulse must in the end lead to the downfall of any civilization so afflicted lies in its biological undermining. For civilization affords an excellent chance to breed for many sorts of the weak, the stupid, and the vicious characters which are continually arising. . . . It might at first sight appear that this can lead to a situation little or not at all worse than that existing at present, so long as the sound and the superior also reproduce, mix with, and counterbalance the others; but this impression is erroneous and betrays insufficient understanding of the findings of modern genetics. The crux of the matter is that new mutations, though rare, are continually occurring, and cannot be prevented, and that, in any population whatever, the mutations which give rise to 'defective' or 'pathological' traits are relatively far more abundant than the 'beneficial' mutations. We cannot, therefore, by any possibility escape the conclusion that the process of mutation will in time cause a gradual heaping up of undesirable traits of all sorts in any group of animals or plants, if individuals having the defective genes are allowed to multiply merely at the same rate as the others. One after another the remaining 'normal' genes will themselves 'go bad' through mutation. . . . So long as this condition holds, then, a biological disorganization will take place, and will necessarily continue without limit, or until the species disappears. Actually, there are indications that under civilization many traits that are detrimental to the species as a whole are propagated at a *faster* rate than the normal, so that the otherwise exceedingly slow biological degeneration is thereby hastened. This, when the increasing complexities of living are really calling for even greater ability on the part of the group as a whole." —Muller, *Out of the Night, op. cit.,* pp. 42-43.

7 "There can be no manner of doubt that in civilized society the weak, the deformed, the foolish, the insane, and degenerates of all kinds, have a much greater opportunity to survive and propagate their defects than they commonly had among primitive peoples." —S. J. Holmes, *Studies in Evolution* (New York: Harcourt, 1923), p. 69.

offspring of the more intelligent. As a result, we have today reached a condition in which those of mediocre intelligence[8] have come to out-number all others to a greater extent than ever before.[9] There is strong evidence to be presented that their preponderance increases daily, causing the *proportion* of good minds in the population continually to diminish.[10]

It is not to be expected that this diluting influence can go on,

––––––––

8 A person of mediocre or average intelligence (I.Q. = 100):

Has difficulty perceiving a similarity between egg and seed.
Confuses the distinction between ability and achievement.
Has considerable difficulty with a problem of the sort: If a man buys eight 9¢ cigars, how much change should he get back from a dollar? (No tax involved).
Frequently does not understand the notion that an event is caused, not by a single agent, but by a number of factors acting over time.
Cannot figure the monthly interest paid on a loan charged at a given yearly rate.

Today the *majority* of people cannot cope with these.
9 This statement means exactly what it says. If the proportion of intelligent individuals in the population is not increasing as rapidly as the population in general, i.e., if it is not doubling every 33 years, then that proportion is decreasing. ("The population of the world is now doubling in 33 years." —Alan Guttmacher, in address to L.A. World Affairs Council, April 12, 1965.) Every evidence shows that, at least in the late and highly urbanized stages of civilizations, the most intelligent segments reproduce less rapidly than the generality. Thus they become a diminishing proportion of the whole while, during the million and more years of the hunting stage, they represented the preponderance of the adult population.
To the psychometrist this would mean that the I. Q. distribution curve is becoming more leptokurtic and reduced in standard deviation. Both Burt and Cattell have discussed this possibility, but the short time intervals over which populations have been re-tested so far do not as yet permit a definite conclusion. However, the anatomical evidence from paleontology suggests the same trend, but on a far greater scale.
10 "In the case of mental traits even more than of those having to do with general health and vigor, there is reason to conclude that selection has greatly relaxed under modern conditions." —Hermann J. Muller, "Our Load of Mutations," *American Journal of Human Genetics*, Vol. 2, No. 2, June, 1950, p. 165.
This situation creates relatively rapid biologic effects when it comes to intelligence, due to the fact that characteristics most recently acquired in our philogenetic development are most quickly lost when selection is weakened. That is, reversing mutations occur most frequently in the less firmly established traits such as intelligence.

generation after generation, without effect upon the stock in which it occurs. It is like the encroachment of weeds upon a garden once well tended. Its effect is difficult to discern in man within a generation and so it tends to be overlooked.[11] Yet the

------

11 "Burt,[a] Cattell[b] and, indirectly, Lentz[c] and Willoughby,[d] have presented evidence of a decline in intelligence quotient (I.Q). Tuddenham[e] and Hunt[f] have presented evidence of some increase but Burt and Cattell present evidence that this increase is only in crystalized intelligence; i.e. in that part of intelligence test performance which corresponds to school achievement, and which represents improvement in education. Higgins, Reed[g] and Bajema[h] have pointed to the possibility of a faint increase in I.Q. in certain populations, although they do not take account of the illegitimate birth rates or of the non-white birth rates in those populations. Also, greater weight must be given to the Cattell findings than to either those of Bajema and Higgins, Reed and Reed because: 1) Cattell's sample was far larger; 2) it includes an actual repeat testing of the population after 13 years and 3) it used the new Culture Fair intelligence tests so as to avoid spurious effects from educational contamination. Thomson[i] also has pointed to the possibility of such an increase (in Scottish children), although he regards this as an artifact. None mention a different yet related phenomenon: the rise of birth rate among the mentally deranged.[j] None compare the birth rates of the segments of populations studied with the birth rate of the world population.[k]

What is emphasized in this book is not so much a drop in I.Q. as an increasing preponderance of the mediocre—a burgeoning of the masses—throughout much of the world. This great heaping up about the midpoint of the I.Q. distribution curve does not of itself produce a decline in average I.Q., but does produce the decreasing proportion of the intelligent of which we speak. (Fig. 8).

[a]Sir Cyril Lodowic Burt, *Intelligence and Fertility* (London: Hamish Hamilton, Ltd., 1946).

[b]Raymond B. Cattell, *The Fight for Our National Intelligence* (London: King, 1937), p. 4.

[c]T. F. Lentz, "The Relationship of Intelligence Quotient to Size of Family," *Journal of Educational Psychology*, Vol. 18, pp. 486-496, 1927.

[d]R. R. Willoughby, "Fertility and Parental Intelligence," *American Journal of Psychology*, Vol. 40, pp. 671-672, 1928.

[e]Tuddenham, R. D., "Soldier Intelligence in World Wars I and II," *American Psychologist*, Vol. 3, pp. 54-56, 1948.

[f]J. McV Hunt, *Intelligence and Experience* (Ronald Press: New York, 1961).

[g]J. V. Higgins, E. W. Reed and S. C. Reed, "Intelligence and Family Size," *Eugenics Quarterly*, Vol. 9, 1967, pp. 84-90. Also: Sheldon C. Reed, "The Evolution of Human Intelligence," *American Scientist*, Vol. 53, 1965, pp. 312-26.

[h]Carl Jay Bajema, "Estimation of the Direction and Intensity of Natural Selection in Relation to Human Intelligence," *Eugenics Quar-*

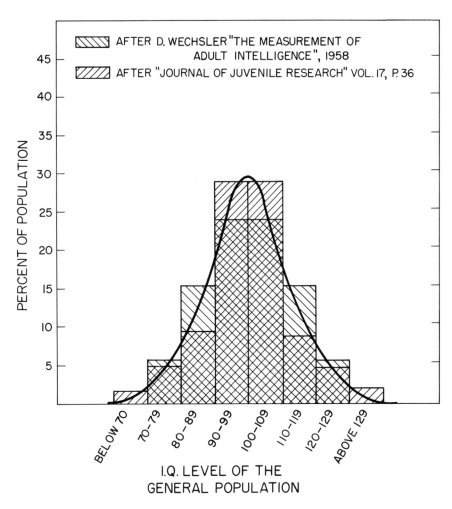

Fig. 8.   This graph shows the great preponderance today of those with average intelligence.

(The technical aspects of intelligence distribution are described in *Abilities: Their Structure, Growth and Action* by Raymond B. Cattell (Boston, Houghton Mifflin, 1970.)

effect is there and it is being intensified at a compound rate. It could not be otherwise unless man were an exception to biologic laws.[12] Quantitatively, man is an enormous success. Yet qualitatively he appears to be, on the average, retrogressing. Disturbing though the realization is, the fact remains that man, the lord of the earth, is allowing himself to become a decadent thing. Much the same result could be expected of any type of creature which, freed from natural selection, continued to breed indiscriminately. This problem of degradation is the most fundamental one with which all improvers of plants and animals contend. That it can be contended with successfully gives rise to the greatest hope of man. But before the hope can begin to be realized, we must recognize—so as successfully to combat—some of the imme-

---

terly, Vol. 10, No. 4, 1963, pp. 175-87. Also, *Eugenics Quarterly*, Vol. 13, 1966, pp. 306-15; and Vol. 15, No. 3, 1968, p. 198.

(Cattell and Bajema wish it to be pointed out that one of the possible sources of difference between their findings lies in the fact that Cattell determined the relation of intelligence of children to the number of children in the family, whereas Bajema determined the relation of intelligence of parents to the number of their children.)

[i] *The Trend of Scottish Intelligence* (London: University of London Press, Scottish Council for Research in Education, 1949). Also:

Godfrey Thomson, *The Trend of National Intelligence*, Eugenics Society Galton Lecture (London: Hamilton, 1947).

[j] In one study the reproductive rate of schizophrenic women was found to have increased 86 percent, while that of the general population increased 25 percent. —Franz J. Kallman, *Medical Tribune*, April 3, 1964, p. 22.

[k] The present world birth rate is 34 per 1000. (It is 39 to 50 per 1000 in the less developed countries; 17 to 23 per 1000 in the more developed countries.) —U. N. *Demographic Yearbook*, 1965.

12 The second law of thermodynamics expresses the tendency of all systems to run down. (Harrison, *What Man May Be, op. cit.*, p. 68.) Life itself is no exception. It too tends to run down and the predominance of adverse mutations is the means by which it degrades itself.

Yet organic evolution has shown in general a tendency toward more, not less, complex systems and it has been an accelerating process. It is chiefly the force of natural selection, tending to choose for survival the advantageous mutations—the beneficial rare exceptions to the rule—which has resulted in life not running down but repeatedly transcending itself. Natural selection, when sufficiently rigorous, negates the operation of the second law. When it is not sufficiently rigorous, the second law may operate inexorably.

It is against the background of such basic forces that we must consider the consequences of man's present interference with natural selection and his failure, so far, to provide for himself a logical successor to it.

diate and major results of the present breeding characteristics of man. Part II of this book is concerned with this situation.

2.   The Decreased Reproduction of the Intelligent

". . . primitive man . . . did not yet recognize the connection between the sexual act and conception."[13]

Before human beings realized that pregnancy followed sexual union, presumably almost all took part in such union as often as circumstances and their own desires dictated. That is still true of aborigines who even today have not perceived the correlation.

However, with the domestication of grazing animals several thousands of years ago, men and women could observe and control the activities of their animals, including their sexual activities. Probably as a result of this, someone deduced what causes females to conceive.

The realization of the cause of pregnancy led directly to ways to forestall pregnancy in humans. The most obvious techniques were abstention, continence and celibacy. Later it was realized that these Spartan approaches were not necessary, provided spermatozoa were prevented from meeting ova. Other ways, including abortion, also came to be practiced. As information about how to impede pregnancy spread among humans, those with the intelligence and self-control to make the most effective use of it were the ones most able to reduce the number of their children below what would naturally have been born to them. This had eventually a tremendous influence on the quality of man. It added to the abnormal increase of the less intelligent an artificial reduction in the numbers of the more intelligent.[14]

------

13 Herbert Wendt, *In Search of Adam* (Boston: Houghton Mifflin, 1955), p. 363.

14 The greater their capacity for understanding, the better the individuals understood how to limit the number of their children. How many less than the natural number the majority of intelligent women have today is indicated by the fact that "a normal woman whose childbearing is not restricted will have an average of 15 live children." —Alan Guttmacher.

Physicians and teachers are among the categories with highest intelligence. Yet, "Teachers and doctors, as is well-known, are the most rapidly dying sections of the community. They average decidedly less

"The tendency of civilized life to sterilize its ablest citizens . . . is the experience of nearly all countries which enjoy even a passable degree of prosperity. It is perhaps more marked now than ever before, but it has certainly occurred at other periods of history. For example, the earlier Roman emperors were continually in difficulty because of the extinction of the senatorial families, which were the class whose administrative ability had been so largely responsible for the creation of the Roman Empire."[15]

The ancient Egyptians were familiar with anti-conceptive methods. Abstention from intercourse during the days of ovulation, which has received much attention of recent years, was taught in ancient Greece.[16] Infibulation (mechanical prevention of copulation) was well known in ancient Rome.[17] The condom was used in ancient Rome and vaginal douches also. The sexual impulse and reproduction were recognized as distinct in the ancient, defunct empires, as they are today.

Contraceptive measures were not so efficient in earlier days as now but they were employed, particularly in the highly urbanized stages of civilizations. They are by no means a recent devisement, although their use and the very phrase, "birth-control"

---

than two children per pair of adults." —Raymond B. Cattell, *The Fight for Our National Intelligence* (London: King, 1937), p. 118.

"Despite their low mortality, the intellectual classes of most countries do not represent a self-perpetuating group. And, despite their higher mortality, the stocks on the lower intellectual levels show the greatest preponderance of births over deaths." —Samuel J. Holmes, *The Eugenic Predicament* (New York: Macmillan, 1933), p. 116.

"Not only are college graduates themselves failing to reproduce proportionately, but all classes whose intelligence enables them to attain positions of responsibility in the business and professional worlds are reproducing less than their proportionate quota.

"Inevitably it happens, then, that the children of each generation are being drawn in greater proportion from the lower intellectual strata of the population." —Laurance H. Snyder, *The Principles of Heredity* (Boston: D. C. Heath and Co., 1935), p. 339.

15 Sir Charles Galton Darwin, *The Next Million Years* (Garden City: Doubleday, 1952), p. 91.

16 Richard Lewinsohn, *A History of Sexual Customs* (New York: Harper, 1958), p. 322.

17 *Ibid.*, p. 324.

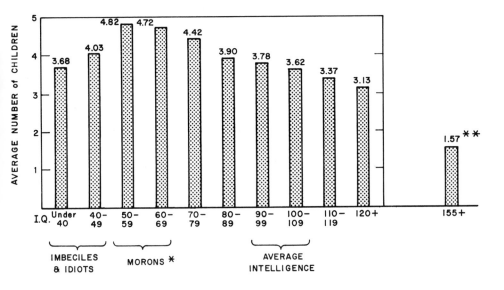

Fig. 9.   FAMILY SIZE IN RELATION TO I.Q.'S OF CHILDREN

The boxed chart is from W. F. Pauli, *World of Life* (New York: Houghton Mifflin, 1949, Figure 522.)

*This table indicates that the greatest number of births is among morons instead of among the average. It is probable that the death rate among the children of morons is also higher than among the average, resulting in predominance among adults of the present average (I.Q. = 90-109). Some natural selection still operates.

**Mensa is a worldwide organization limited to the most intelligent two per centum of the population. When the number of children of every member of this group past the child-producing years (age 45) is tabulated from the 1963-64 Mensa Register, their average is found to be only 1.57 children per couple.

It is recognized that the two sources of information used in the graph above are not strictly comparable. However, they are entirely consistent.

have had relatively recent popularization.[18] Due to increased effectiveness of birth control methods, many of the intelligent have become in this century more skilled and sophisticated at evading pregnancies than ever before.[19]

"*. . . at least one half of the children who would prove the most effective and most valuable citizens . . . are annually withheld from us.*"[20]

At the same time the less intelligent, being less interested in birth control, or less effective in their efforts to employ it, continue to multiply more nearly at the natural rate.[21]

I set it down that the day man learned the cause and prevention of pregnancy was one of the darkest in the evolution of humankind so far. This was the day when there was opened a

---

18 "The event that really launched the ships was the Bradlaugh-Besant trial of 1877-8. Charles Bradlaugh and Mrs. Besant reprinted Dr. Charles Knowlton's pamphlet on birth control called The Fruits of Philosophy . . . they also informed the police, and in due course were arrested. . . .

"It was after this trial and the publicity which followed, . . . that the birth rate in Great Britain, then standing at 35 per thousand, began to fall, and it went on falling until it reached its lowest peace-time level of 14.9 in 1933." —Margaret Pyke, "Family Planning," *Eugenics Review,* Vol. 55, No. 2, July, 1963, p. 71.

Let no one think that attitudes about births cannot change the very size and character of nations.

19 "In the past there has been considerable selection in favor of intelligence characteristics involving abilities to learn, to solve problems, and to transmit experience to offspring. In recent decades this pattern of selection has been completely reversed. Whereas in former times high intelligence increased the probability that many of an individual's characteristics would be reproduced and would spread throughout the population, today a high intelligence actually decreases this probability. The present situation has arisen as a result of the uneven adoption of birth-control techniques by differing social and economic groups in the Western World." —Harrison Brown, *The Challenge of Man's Future* (New York: Viking, 1954), p. 102.

20 W. C. D. and C. D. Whetham, "Decadence and Civilization," *Hibbert Journal,* Vol. 10, No. 1, October, 1911, p. 183.

21 It is pointed out by Sam Flanz, M.D., that intelligence makes possible to its possessors many engrossing interests in addition to sex, whereas those with more limited minds find sex one of their main interests. To the extent that this is so, the intelligent have a constitutional disadvantage with respect to the others, now that birth-rate selection has largely replaced natural selection. This factor alone would suffice to produce a decline in relative numbers of the intelligent.

way to diminish the renewal of the intelligence which natural selection had for millions of years been building in man and his precursors. Artificial selection by differences in birth rate began that day to oppose natural selection by differences in death rate as a shaping force. Ever since, this type of artificial selection has been almost entirely adverse in its effect on the intelligence of man. Birth control may someday save us from worldwide starvation; it may even come to be used in such a way that man can surpass his present self. But until now it has been almost wholly pernicious in its effect upon man.[22]

"The most intelligent individuals, on the average, breed least, and do not breed enough to keep their numbers constant. Unless new incentives are discovered to induce them to breed, they will soon not be sufficiently numerous to supply the intelligence needed for maintaining a highly technical and elaborate system— meanwhile, we must expect that each generation will be congenitally stupider than its predecessor. This is a grave prospect."[23]

The trend toward the practice of birth control was, and is, stronger in some people than in others; within certain religious groups than in others; at some periods of time than others.[24] Birth

---

22 ". . . a number of signs discernible to the sociologist have pointed to the grim conclusion that we have entered on a phase in which low-grade mentality is reproducing itself with unhindered prolificness, whilst families of more than average intelligence are failing even to maintain their numbers." —Cattell, *The Fight for Our National Intelligence, op. cit.,* p. 4. Brazil is said to represent an exception to this. (See John B. Griffing, "A Comparison of the Effects of Certain Socio-Economic Factors—," *Journal of Heredity* XXXI 1940, pp. 13-16.) Utah may also be an exception.

23 Charles Beard, *Whither Mankind* (London: Longman's, Green and Co., 1934), p. 80.

24 Exceptionally capable peoples, living under an economic system which was chiefly agricultural (as pre-industrial England, U. S., etc.) where children were an economic advantage, at times *increased* the numbers of intelligent offspring markedly. For example, during the first 200 years of the American Colonies and the young U. S. A. the average number of children born was 8 per woman.

"The original number of persons which had settled in the four provinces of New England in 1643 was 21,200. Afterwards it was calculated that more left them than went to them. In the year 1760 they were increased to half a million. They had, therefore, all along doubled their number in 25 years." —T. R. Malthus, *The Principle of Population,*

control among the intelligent seems to be practiced most intensively during the late stages of civilizations.[25] Indeed, it seems
likely that the practice contributed substantially to the decline
and fall of many civilizations.[26] Although it is unnatural to have

7th edition, (substantially the same as 2nd edition, 1803), Vol. I (London: J. M. Dent and Sons, 1914), pp. 305-6.

"The highest reproductive rates known for man (today) are about
50 births per thousand population per year. The figure for the United
States (1620 to 1820), when the population was going through as rapid
expansion as any ever observed, has been calculated at 55 per thousand."
—Marston Bates, *The Prevalence of People* (New York: Scribner's, 1955).
p. 92.

In England there are indications that, until at least the middle of the
nineteenth century, marriage was less frequent in the lower classes and
child survival was higher in the upper classes. These were the times when
the human bases of the greatness of the English speaking peoples were
being built and expanded. These were the times when they came to
occupy most of North America, Australia, South Africa, New Zealand,
etc.

25 Intelligence is not required in order to have children. The lowliest
animals can accomplish this. But intelligence has been required to *avoid*
having children. For this reason, among others, the effect of birth avoidance to date has been selectively to reduce the numbers of the intelligent.

"Nor can there be any question that intelligence has in part a genetic
basis, and it is quite immaterial for our argument whether the heritability
of intelligence is 25 percent or 75 percent. Finally, there is abundant
statistical evidence that in most communities those people whose professions require high intelligence produce on the average smaller families
and at a later age than do people like unskilled laborers, whose professions
do not make such requirements. Even though it is still being heatedly
denied by identicists, the weight of the available evidence fully supports
Huxley's conclusion that those who are intellectually best endowed contribute less to the gene pool of the next generation than do the average
and, indeed, most of the less than average." —Mayr, *Animal Species and
Evolution, op. cit.,* p. 659.

". . . it does appear that the feeble-minded, the morons, the dull and
backward, and the lower-than-average persons in our society are out-
breeding the superior ones at the present time. Indeed, it has been estimated that the average Intelligence Quotient of Western population as a
whole is probably decreasing significantly with each succeeding generation." —Brown, *The Challenge of Man's Future, op. cit.,* pp. 102-3.

26 "Civilization tends to extinguish its best stock and thus to impoverish its racial inheritance. As fast as the hereditary factors for
superior mentality combine and manifest themselves in individuals of
distinction, they tend to disappear. These factors are the most precious
possession of the human race, and they are undergoing a heavy drain. All
this is brought about by the simple fact that intelligence has discovered
the means of out-witting nature by sacrificing posterity to present welfare." —Holmes, *Studies in Evolution and Eugenics, op. cit.,* p. 133.

(Continued on next page)

small families, it became at times actually stylish to do so. This proved to be the consummate folly of its practitioners—the step leading to the decline or destruction of the classes and civiliza-

---

"Babylonia, Egypt, Greece and Rome did not fail for lack of children, but for a dearth of children of the right kind. There is no doubt that civilization exposes the ablest of her votaries to an extreme temptation—the temptation of luxury. The basal and essential condition for racial progress is a healthy birth rate—in every grade of every community. As we raise our standard of living, the more expensive becomes the maintenance and education of our children; we become more and more tempted to spend on ourselves what should be sacrificed in raising another generation." —Keith, *Concerning Man's Origin, op. cit.,* p. 125.

"Not only had the increase of intelligence been the basis for civilizations, but decline of intelligence has been the basis for the disintegration of civilizations. Evolution leads to the amplification of intelligence, and by way of intelligence to a civilization, but civilization normally causes a decline of intelligence by terminating the weeding-out process, thus permitting the higher birth rates of burdensome people to set the direction of change." —Pendell, *Sex vs. Civilization, op. cit.,* p. 218.

"Problem-makers reproduce in greater percentage than problem-solvers, and in so doing cause the decline of civilizations." —Pendell, *Sex vs. Civilization, op. cit.,* p. 138.

"Generally speaking, throughout the Middle Ages, so far as is known, the so-called upper classes contributed more to succeeding generations than the so-called lower classes. (In part because the death rate among the lower classes was then much higher than among the upper classes.) The position is now reversed, and it would appear that in the later days of Greece and Rome also the upper classes contributed less to the population than the lower classes. It is possible that the same may have been the case in other ancient empires. The utmost importance has been attributed to the influence of this form of differential fertility upon the course of history. Mr. McDougall, for instance, writes as follows: 'Looking at the course of history widely, we may see in the tendency of the upper strata to fail to reproduce themselves an explanation of the cyclic course of civilization.'" —William McDougall, *Group Mind* (London: Cambridge, 1939), p. 260. Quoted by T. H. Marshall, Sir Alexander Carr-Saunders, and others, *The Population Problem . . .* (London: Allen, 1938), p. 457.

". . . it is probable that the fall in the birth rate among the . . . abler stocks, together with the constant drain of incessant wars . . . not only killed off many of the best in each generation, but also, by the survival of the unfittest, lowered the average quality of the nations. Doubtless the obvious military and other causes, usually blamed, had much to do with the catastrophe, but economic and racial factors must not be overlooked. We may perhaps say that the overthrow of Rome by the Northern invaders was not so much a destruction of civilization by barbarians, as the clearing away of a doomed and crumbling ruin. . . ." —Sir William Dampier, *A History of Science* (London: Cambridge University Press, 1966), p. 69.

tions which practiced it. As the intelligent let themselves die out, their way of life died with them, or they were overwhelmed by more vigorous peoples, nations or civilizations.

3. The Cessation of Selection For Intelligence

Not only do the intelligent ones fail on the whole to maintain their former and natural predominance in the population, there is no longer any selective advantage to those rare mutations in the direction of even higher intelligence.[27] Those with the greatest minds, some of which may represent not just a fortunate combination of genes for high intelligence, but a genuine mutational increase in brain capacity such as occurred and throve time and again in the primitive eras, no longer have the greatest number of living offspring. No longer do the progeny of such supplant the progeny of lesser men. Selective pressure toward higher intelligence has practically ceased.

4. The Acceleration of the Mutation Rate

Another adverse influence on the nature of man has but recently been intensified by advances in physical and chemical science. Each exposure of the reproductive organs to x-rays, or to radioactivity, or to hallucinatory drugs such as LSD, increases the occurrence of mutations in subsequent offspring,[28] and almost all mutations are harmful.[29]

The rate at which this new effect is damaging the race of man is a matter widely debated today. We cannot hope to arrive at a figure here. But practically no one maintains that the effect is not damaging.

"Despite disagreement in details among geneticists, the con-

---

27 ". . . there is no selection encouraging favorable variations." —Huxley, *Man in the Modern World, op. cit.,* p. 60.

28 *The Biological Effects of Atomic Radiation* (Washington, D.C.: National Academy of Sciences, National Research Council, 1956). Also H. J. Muller, "How Radiation Changes the Genetic Constitution," *Bulletin of Atomic Scientists,* 1955, 11:329-338, 352.

29 ". . . it should be borne in mind that the great majority (ordinarily well over 99 percent) of induced mutations, as of spontaneous ones, are of a detrimental kind. . . ." —H. J. Muller, "The Prospects of Genetic Change," *American Scientist,* Vol. 47, No. 4, December, 1959, p. 553.

clusion that an increased mutation rate is harmful is universally accepted."[30]

Since we evolved upward through mutations, we should evolve upward all the more rapidly with an increased mutation rate,[31] except that we no longer have adequate selection, natural or otherwise, to preserve the advantageous mutations and eliminate the disadvantageous.[32] Since an increase in the mutation rate is, under present circumstances, harmful to man, it follows that the basic mutation rate—favorable for millions of years—is also harmful to him now.[33] We should be able to note evidences of harm to the general constitution of man if we are willing to recognize them. And we need to be willing. The evidence, if sufficiently recognized, can lead to changing harmful influences into constructive ones.

The four sinister and pervasive influences mentioned here combine to produce in each generation of men an ever greater proportion of the mediocre than existed in previous generations. At present most of the intelligent classes are more widely addicted to birth control than ever before; the masses are multiplying and surviving at a rate almost beyond comprehension and there is a grave question as to how long man can maintain his present high civilization despite a growing burden of inherent deficiences.[34] Man is wasting a magnificent inheritance.[35] He who

---

30 James F. Crow, "Mechanisms and Trends in Human Evolution," *Daedalus*, Summer, 1961, p. 426.

31 I am indebted to Professor Raymond Cattell for this perceptive observation.

32 "Mutation alone, uncontrolled by natural selection, could only result in degeneration, decay, and extinction." —Theodosius Dobzhansky, *The Biology of Ultimate Concern* (New York: New American Library, 1967), p. 41.

33 "Geneticists estimate that a minimum of . . . 20% of the progeny of humans in every generation carry newly arisen more or less harmful mutants." —Frederick Osborn, *The Future of Human Heredity* (New York: Weybright and Talley, 1968), p. 76. Also Theodosius Dobzhansky, *Mankind Evolving* (New Haven: Yale University Press, 1962), p. 50.

34 The better the communications which exist, the more can a small percentage of intelligent ones coordinate their efforts and so advance knowledge and technology if they will. Thus scientific progress can still occur, even if the proportion of the intelligent recedes. Also, with population rapidly expanding, there can be an increase in the actual num-

has come to control the natures of living things about him has yet to accomplish the basic control of his own nature. As a consequence the greatest problem of man is no longer his natural enemies. Now it is man himself.

---

ber of those able to devise advances, even as their proportion in the population declines. Thus improved communications and swiftly growing numbers can help conceal a drop in the proportion of the intelligent, up to the point of breakdown in communications. The iron curtain, though not complete, is one such breakdown.

35 ". . . I am unhappy that the pool of human germ plasm, which determines the nature of the human race, is deteriorating. The collection of molecules of deoxyribosenucleic acid that will make the next generation of human beings what it will be is not so good as that which determined our character; there are more bad molecules in the collection. The defective genes are now not being eliminated from the pool of human germ plasm so rapidly as in the past, because we have made medical progress and have developed feelings of compassion such as to make it possible for us to permit the individuals who carry the bad genes to have more progeny than in the past. Moreover, defective genes are being manufactured at a greater rate than in the past, because there are new mutagenic agents operating in the world today." —Linus Pauling, "Molecular Disease," *American Journal of Orthopsychiatry*, Vol. XXIX, No. 4, 1959, p. 684.

". . . the general quality of the world's population is not very high, is beginning to deteriorate, and should and could be improved. It is deteriorating, thanks to genetic defectives, who would otherwise have died, being kept alive, and thanks to the crop of new mutations. . . . In modern man the direction of genetic evolution has started to change its sign, from positive to negative, from advance to retreat; we must manage to put it back on its age-old course of positive improvement." —Sir Julian Huxley, "The Future of Man—Evolutionary Aspects," in *Man and His Future*, Gordon Wolstenholme, ed. (Boston: Little Brown and Co., 1963), p. 17.

"Man's genetic nature has degenerated and is still doing so. . . . It seems now to be established that, both in communist Russia and in most capitalist countries, people with higher intelligence have, on the average, a lower reproductive rate than the less intelligent; . . . the genetic differences are slight, but . . . such slight differences speedily multiply to produce large effects. If this process were to continue, the results would be extremely grave." —Mayr, *Animal Species and Evolution, op. cit.,* p. 658, quoting Huxley, *Evolution in Action, op. cit.*

We can face this fact with the happy-go-lucky attitude of the profligate that "we aren't bankrupt yet," or with that of a capable administrator who perceives the leakages which can founder an enterprise, and who corrects them before they do so.

PART II

NOW

CHAPTER I

MAJOR CONSEQUENCE OF THE INCREASE OF THE LESS INTELLIGENT

Throughout the million or more years of the hunting stage, intelligent men often dispatched their immediate enemies, human or otherwise, else they would not have survived to become our forebears. Yet toward their own people, intelligent or not, they were (and are) largely acceptive. So long as those around the intelligent were not overt enemies, whether they were relatives or not, fit or unfit, they were seldom deliberately exterminated.[1] Thus, as the rigors of natural selection were progressively lessened, more and more of the less intelligent were enabled to survive and reproduce along with the intelligent. With the discovery of agriculture a markedly greater proportion of the less intelligent began to survive. This proportion has increased even more rapidly in the past 250 years of the Industrial Revolution. Today, in nations where figures are available, about 70 percent of the population possesses mediocre intelligence or less.[2] Huge masses of those with ordinary minds have been enabled to come into being and proliferate.

When these masses preponderate sufficiently to give them the upper hand, a live-and-let-live acceptance no longer prevails. As the proportion of those with mediocre intelligence increases, there occur certain ominous manifestations of its growing prevalence and power. In areas where the proportion becomes great enough, and the numbers and leadership of the other classes become sufficiently weakened, a potentially explosive situation is created. If incendiary leaders arise[3] to ignite this explosive atmos-

----

1 Exceptions to this were societies which practiced selective infanticide, such as the Spartan and Tahitian.

2 Fig. 8 shows that about 70 percent of today's population possesses an I.Q. below 105. (See page 87.)

3 There are almost always opportunists who in times of unrest will betray their own class in order to attempt to wield the power of the masses.

101

phere, social detonation occurs. *The masses are incited to turn upon the "haves" of their very own people and slaughter them.*

Here emerges a thing so basic in its nature and tremendous in its effects that it has great historic significance. More than that, it exerts today major evolutionary effects upon man himself. An examination of the phenomenon is gruesome, but it is essential to an understanding of the human situation, for today mass revolutions—class wars—exceed in deadliness the most lethal international wars of all time.[4]

The masses kill so as to seize what others possess. They also kill so as to remove galling evidences of their own relative inferiority. They are impelled to obliterate all whom they feel to be above them, whether superior in wealth, position, character, beauty or intellect. Of the intelligent, only a few of the masses' own most radical leaders are exempted from obliteration in times of mass Terrors.

An example of this vast internal genocide occurred in the French Revolution. It also took place and is still taking place in

----

4 Total slain in the two deadliest wars in history:

Killed in World War I	10,000,000[a]
Killed in Word War II	17,000,000[b]
	————————
	27,000,000

Total slain to 1962 in mass revolutions:

Killed by Communists in U.S.S.R., at least	10,000,000[c]
Killed by Communists in China, at least	30,000,000[d]
	————————
	40,000,000[e]

[a]Sir Charles Darwin, *Problems of World Population* (Cambridge: Cambridge University Press, 1958), p. 25.

[b]Louis L. Snyder, *The War* (New York: Messner, 1960), p. 9.

[c]Admitted by Stalin to Churchill. See Louis Snyder, *The World in the 20th Century* (Princeton: Van Nostrand, 1964) pp. 75-6.

[d]Weyl & Possony, *Geography of Intellect, op. cit.*, p. 147.

[e]One of the latest and most comprehensive reports of this on-going process states: "The approximate number of human beings killed as a direct outcome of the Russian revolution of 1917 and in the implementation of Communist policy is 83,500,000 (excluding World War II casualties). If any reader thinks that this figure is exaggerated, we will willingly supply a breakdown of this total on application." *Intelligence Digest* No. 350, Jan. 1968, p. 11, 41 Rodney Road, Cheltenham, Gloucestershire, England.

the U.S.S.R., in China and in other nations where proletarian revolutions have erupted. Historians have described the manifestations of this appalling aberration. Their words transmit in some measure the nature of the phenomenon. To understand it may be to survive.[5]

THE NATURE OF MASS UPRISINGS

1. Class Warfare in France

The French Revolution (1789-94) is the first great instance of what occurs in a nation when the rabble—the revolutionary segment of the masses—gains the upper hand. The circumstances of this revolution were not concealed behind iron or bamboo curtains, as with today's mass revolutions. Consequently, they permit of some study and justify it.

"The French Revolution . . . revealed itself as different to any revolution that had hitherto been."[6]

The Paris Commune was "the first proletarian government on earth."[7]

"A new and terrible thing has come into the world, an immense new sort of revolution whose toughest agents are the least literate and most vulgar classes, while they are incited and their laws written by (a few) intellectuals."[8]

"The French Revolution was a universal insurrection of the lower orders against the higher. It was sufficient to put a man's life in danger, to expose his estate to confiscation, and his family to banishment, that he was, from any cause, elevated above the populace. The gifts of nature, destined to please or bless mankind, the splendour of genius, the powers of thought, the graces of beauty, were as fatal to their possessors as the adventitious advantages of fortune or the invidious distinctions of rank. 'Liberty and Equality' was the universal cry of the revolutionary party.

---

5 "A threat understood is a threat half conquered."   —Elliott, *The Shape of Intelligence, op. cit.,* p. 231. On the other hand "those who will not learn from the mistakes of the past are condemned to repeat them." —George Santayana, *Life of Reason,* Vol. I (New York: Scribner's, 1932), p. 284.

6 *Encyclopedia Britannica,* 14th ed., Vol. 9, 1937, p. 804.

7 Karl Marx.

8 Alexis de Tocqueville, *The European Revolution,* 1st ed., 1865 (Garden City: Doubleday Anchor edition, 1959), p. 161.

Their liberty consisted in the general spoliation of the opulent classes; their equality in the destruction of all who outshone them in talent or exceeded them in acquirement."[9]

"The vengeance of the tyrants fell with peculiar severity upon all whose talents or descent distinguished them from the rest of mankind. . . . Lavoisier was cut off in the midst of his profound chemical researches; he pleaded in vain for a respite to complete a scientific discovery. Almost all the members of the French Academy were in jail. . . ."[10]

"The combination of wicked men who thereafter governed France, is without parallel in the history of the world. Their power, based on the organized weight of the multitude . . . was irresistible. By them opulent cities were overturned; hundreds of thousands of deluded artisans reduced to beggary; agriculture, commerce, the arts destroyed; the foundations of every species of property shaken. . . . All bowed the neck before this gigantic assemblage of wickedness. . . . There was no medium between taking a part in these atrocities, and falling a victim to them."[11]

"Five hundred thousand persons, drawn from the dregs of society, disposed . . . of the lives and liberties of every man in France."[12]

Many historians have described the bloody occurrences which took place in France. Brief accounts of some of them follow:

The Massacres of the Priests

The rabble turned upon the priests.

"After the 10th of August, they had shut up . . . relatives of émigrés, journalists, priests, aristocrats of all ranks. . . .

"Towards 3 o'clock in the afternoon of September 2, 1792, the suspects who had been packed into the gaol heard a distant noise ascending the street. . . . A dense mob, clamourous, shouting, wildly excited, debouched from the Rue du Buci. On it came. The sight of the prisoners clinging to the bars of the prison drew

---

9 Sir Archibald Alison, *History of Europe*, Vol. I (Edinburgh and London: Wm. Blackwood & Sons, 1849), p. 52.

10 *Ibid.*, Vol. III, pp. 302-303.

11 *Ibid.*, Vol. III, pp. 245-246.

12 *Ibid.*, Vol. III, p. 150.

from the populace a dreadful yell of fury. Ten thousand threatening fists were raised.

"You saw emerge then [from a carriage arriving at the prison] a tall young man clad in a white gown. He stretched out his arms towards the crowd; he turned to right and left and cried: 'Mercy Mercy! . . . Pardon!'

"These words aroused the populace, who began to howl ferociously. There was a scuffle . . . ten sabres descended upon the young priest; a long red stain appeared upon his white garment, and, gently he fell. Other priests could be perceived, huddled together, within, pale and dumb with terror. . . . The mob made a rush, and, of the twenty-four priests . . . twenty-two were butchered on the spot."[13]

"In such fashion did the September Massacres begin!"[14]

". . . the slaughter lasted for fourteen hours. When the dawn broke after this terrible night . . . the total number of the dead was 270."[15]

". . . the mob . . . butchered all the ecclesiatics, who, they said, had been put into the fold there."[16]

"The butchery went on at the other prisons on the following days. . . . The murderers became so intoxicated with slaughter that common-law and political prisoners, women and children, were slain indiscriminately. Some of the bodies . . . were horribly mutilated. There are different estimates of the numbers of the slain, varying from 1,100 to 1,400. The populace looked on indifferently or with satisfaction at these scenes of horror.

"Those who . . . furnished most of the victims (of the September Massacres) were everywhere the . . . priests."[17]

The Communist Terror in Paris

"Give me 300,000 heads; hang above their doors all mer-

---

13 Louis Leon Theodore Gosselin, *Paris in the Revolution* (New York: Brentano's, 1925), pp. 111-12.

14 *Ibid.,* p. 123.

15 *Ibid.,* p. 126.

16 Edward Leroy Higgins, *French Revolution* (Boston: Houghton Mifflin, 1939), p. 250.

17 Albert Mathiez, *The French Revolution* (New York: Russell, 1956), pp. 181-84.

chants, bakers and grocers and I will guarantee that the country is saved." (Marat, "Friend of the People.")[18]

". . . the movement sprang from the communistic preaching of Jacques Roux and his rivals, Varlet and Leclercq. It was they who launched the idea that the political revolution ought to be completed by a social revolution, and that equality of civil rights would be useless without equality of fortunes. . . . Communism had won the victory, and . . . the tendency towards the Left had not stopped short at the 'bourgeois' Revolution, but went the length of a 'proletarian' Revolution.

"On the 26th of February . . . the Convention decreed . . . that the property of persons hostile to the Republic was to be confiscated. . . . On the 13th of March a decree of the Convention proclaimed as traitors to the country, subject to the death penalty, all those who should excite misgivings as to the food supplies, attempt to corrupt public opinion, or work for a change of government. . . . Finally, on the 10th of June the famous law of 22 Prairial drew up a complete list of the crimes punishable with confiscation of property or death . . . the list was so long that there was not a Frenchman who could not consider himself destined for the guillotine.

". . . the situation was clear. . . . The object was to exterminate three hundred thousand families in order to seize their property . . . it was . . . 'the handing over of the individual, with all his rights, to the community,' in strict accordance with the rule laid down by Rousseau . . . their doctrines had developed, through successive stages, from anarchic Liberalism to Communist dictatorship. (p. 307) As the collective Sovereign . . . they disposed arbitrarily of fortunes, liberties and lives.

"All that was best in France was in hiding or at the front. It was the dregs that governed the country, . . . as Taine says . . . 'Social outcasts and perverts of every sort and description, subordinates full of hate and envy, small shopkeepers in debt, workmen given to loose living and a wandering life, pillars of the coffee-houses, and the drinking-shops, vagabonds of the streets and the countryside, men of the gutter and women of the streets, in short, every kind of male and female anti-social vermin; and

---

18 Quoted by Higgins, *French Revolution, op. cit.,* p. 248.

among this collection a few ranting but sincere demagogues, into whose disordered brains the fashionable theories had found spontaneous entrance; . . . by far the greater number were very beasts of prey, who . . . adopted the revolutionary faith only because it offered them means to sate their appetites.' "[19]

"This period (1793-4) was marked by the height of the Terror . . . the prisons filled faster than they were emptied. Batch after batch was sent to the guillotine in rapid succession. . . . Heads fell like slates from a roof; thirty-one ex-judges of Paris and Toulouse who had once protested against the abolition of the parlements; . . . Lavoisier and twenty-eight farmers-general . . . ."[20]

"Eight thousand prisoners were soon accumulated in the different places of confinement in Paris; the number throughout France exceeded two hundred thousand.[21] . . . From the furthest extremities of France crowds of prisoners daily arrived at the gates of the Conciergerie, which successively sent forth its bands of victims to the scaffold. Grey hairs and youthful forms, countenances blooming with health, and faces worn with suffering, beauty and talent, rank and virtue, were indiscriminately rolled together to the fatal doors. . . . Sixty persons often arrived in a day, and as many were on the following morning sent out to execution. Night and day the cars incessantly discharged victims into the prison.

". . . when the fall of Robespierre put a stop to the murders, arrangements had been made for increasing the daily number to one hundred and fifty. An immense aqueduct, to remove the gore, had been dug from the Seine as far as the Place St. Antoine where latterly the executions took place."[22]

"In removing the prisoners from the jail of the Maison Lazare, one of the women declared herself with child, and on the point of delivery: the hard-hearted jailers compelled her to move on: she did so, uttering piercing shrieks, and at length fell on the ground, and was delivered of an infant in the presence of her persecutors. Such accumulated horrors annihilated all the

---

19 Pierre Gaxotte, *The French Revolution*, Chapter 12: "The Communist Terror" (New York: Scribner's, 1932), pp. 290-317.
20 Mathiez, *The French Revolution, op. cit.*, p. 499.
21 Alison, *History of Europe*, Vol. III, *op. cit., pp.* 262, 265.
22 *Ibid.*, pp. 265-66.

charities and intercourse of life. . . . Passengers hesitated to address
their most intimate friends on meeting; the extent of calamity
had rendered men suspicious even of those they loved the most.
Every one assumed the coarsest dress, and the most squalid appearance; an elegant exterior would have been the certain fore-
runner of destruction. . . . Night came, but with it no diminution
of the anxiety of the people. Every family early assembled its
members; with trembling looks they gazed round the room,
fearful that the very walls might harbour traitors. The sound of
a foot, the stroke of a hammer, a voice in the street, froze all
hearts with horror. If a knock was heard at the door, every one,
in agonized suspense, expected his fate. Unable to endure such
protracted misery, numbers committed suicide."[23]

"Thus, in that frightful delirium which had rendered genius
and virtue and courage suspected, all that was most noble and
most generous in France was perishing either by suicide or by
the blade of the executioner."[24]

"All the distinguished persons confined in the prisons had
fallen; . . . and death was already descending from the upper to
the lower classes of society. We find at this period on the list of
the revolutionary tribunal, tailors, shoemakers, hairdressers,
butchers, farmers, publicans, nay, even labouring men, condemned for sentiments and language held to be counter-revolu-
tionary."[25]

"I have often asked myself how the metropolis of a nation
so celebrated for urbanity and elegance of manners—how the
brilliant city of Paris could contain the savage hordes I that day
beheld, and who so long reigned over it."[26]

The Communist Terror Throughout France

"The whole of the country seemed one vast conflagration
of revolt and vengeance. The shrieks of death were blended with
the yell of the assassin and the laughter of buffoons. . . . Whole
families were led to the scaffold for no other crime than their

---

23 *Ibid.*, pp. 269-71.
24 Louis Adolphe Thiers, *The History of the French Revolution*,
Vol. III (Philadelphia: J. B. Lippincott Co., 1894), p. 223.
25 *Ibid.*, p. 451.
26 *Memoirs of Lavalette*, in Thiers, *French Revolution, ibid.*, p. 374.

relationship; sisters for shedding tears over the death of their brothers . . . wives for lamenting the fate of their husbands . . . and a woman . . . merely for saying as a group were being conducted to slaughter, 'Here is much blood shed for a trifling cause!' "[27]

"Legal murder was the order of the day, a holiday sight, till France became one scene of wild disorder, and the Revolution a stage of blood. The chief actor in this tragic scene, the presiding demon of the storm, was Robespierre."[28]

In Nantes

"In the principal cities of France terror reigned as absolutely as in Paris."[29]

" 'The miserable wretches [at Nantes,] were either slain with poniards in the prisons, or carried out in a vessel, and drowned by wholesale in the Loire. On one occasion, a hundred 'fanatical priests' . . . were taken out together, stripped of their clothes, and precipitated into the waves. . . . Women big with child, children eight, nine, and ten years of age, were thrown together into the stream, on the sides of which, men, armed with sabres, were placed to cut them down, if the waves should throw them undrowned on the shore. . . .' So great was the multitude of captives who were brought in on all sides, that the executioners declared themselves exhausted with fatigue; and a new method of . . . disposing of them was adopted. . . . Two persons of different sexes . . . bereft of every species of dress, were bound together and . . . thrown into the river. . . . It was ascertained by authentic documents that six hundred children had . . . perished by [that] inhuman species of death; and such was the quantity of corpses accumulated in the Loire, that the water of that river became infected. . . . To every representation of the citizens in favor of these innocent victims, Carrier answered, 'They are all vipers; let them be stifled.' Three hundred young women of Nantes were drowned by him in one night. . . .

---

27 Thiers, *The History of the French Revolution, op. cit.*, p. 223. Footnote quoted from Wm. Hazlitt, *Life of Napoleon Buonaparte*, Vol. I (New York: Wiley & Putnam, 1847), pp. 184-5.

28 *Ibid.*, p. 416. Footnote from p. 187 of Hazlitt.

29 *Ibid.*, p. 451.

". . . On another occasion five hundred children of both sexes, the eldest of whom was not fourteen years old, were led out . . . to be shot. The littleness of their stature caused most of the bullets at the first discharge to fly over their heads; they broke their bonds, rushed into the ranks of the executioners, clung round their knees and sought for mercy. But nothing could soften the assassins. They put them to death even when lying at their feet. . . . Fifteen thousand persons perished at Nantes under the hands of the executioner, or of diseases in prison, in one month. The total number of victims of the Reign of Terror in that town exceeded thirty thousand!"[30]

In Lyons

"On October 12 . . . the [Lyons] Convention passed the following motion: . . . 'Every house inhabited by the rich shall be demolished. Nothing shall remain but the poor man's house, . . . buildings specially employed in industry, and public buildings . . .' Collot organized a great . . . celebration, . . . and mass executions began. . . . These butcheries were all the more abominable since they were not excused by the excitement which follows a struggle."[31]

". . . Many women watched for the hour when their husbands were to pass to execution, precipitated themselves upon the chariot, . . . and voluntarily suffered death by their side. Daughters surrendered their honour to save their parents' lives; but the monsters who violated them, adding treachery to crime, led them out to behold the execution of [their relatives!]. Deeming the daily execution of thirty or forty persons too tardy a display of . . . vengeance, Collot-d'Herbois prepared a new and simultaneous mode of punishment. Sixty-four captives of both sexes, were led out at once, tightly bound together, to the Place du Brotteaux; they were arranged in two files, with a deep ditch on each side, which was to be their place of sepulture, while gendarmes, with uplifted sabres, threatened with instant death whoever moved from the position in which he stood. At the extremity of the file, two cannon, loaded with grape were so placed as to

––––––
30 *Ibid.*, pp. 533-34.
31 Mathiez, *The French Revolution, op. cit.*, pp. 400-401.

enfilade the line . . . the signal was given, and the guns were discharged. . . . Broken limbs, torn off by the shot, were scattered in every direction, while the blood flowed in torrents into the ditches on either side of the line. A second and third discharge were insufficient to complete the work of destruction, till at length the gendarmerie, unable to witness such protracted sufferings, rushed in and dispatched the survivors with their sabres. . . . On the following day this bloody scene was renewed on a still greater scale. Two hundred and nine captives . . . were brought before the Revolutionary judges, [and with scarcely a hearing, condemned] to be executed together . . . With such precipitance was the affair conducted, that two commissaries of the prison were led out along with their captives; their cries, their protestations, were alike disregarded. . . . The whole were brought to the place of execution . . . where they were attached to one cord, made fast to trees at stated intervals, with their hands tied behind their backs, and numerous pickets of soldiers disposed so as by one discharge to destroy them all. At a signal given the fusillade commenced; but few were killed; the greater part had only a jaw or a limb broken, and, uttering the most piercing cries, broke loose in their agony from the rope, and were cut down by the gendarmerie. . . . The numbers who survived the discharge rendered the work of destruction a most laborious operation, and several were still breathing on the following day, when their bodies were mingled with quicklime, and cast into a common grave."[32]

In Arras

". . . the representative Joseph Lebon . . . placed on the official black-lists all tax payers assessed at more than 50 francs. 'Considering that among those accused of crimes against the Republic,' he said, 'it is mainly essential to cut off the heads of the rich . . . the criminal tribunal established at Arras will first pass revolutionary judgment on those of the accused who are distinguished by their talents or wealth. . . .' "[33]

"In the city of Arras above two thousand persons . . . per-

---

32 Alison, *History of Europe*, Vol. III, *op. cit.*, pp. 101-12.
33 Gaxotte, *The French Revolution, op. cit.*, pp. 310-11.

ished by the guillotine. . . . Mingling treachery and seduction
with sanguinary oppression, [Lebon] turned the despotic powers
with which he was invested into the means of individual gratification. After having disgraced the wife of a nobleman, who yielded
to his embraces in order to save her husband's life, he put the
man to death before the eyes of his devoted consort. Children
whom he had corrupted were employed by him as spies on their
parents; and so infectious did the cruel example become, that the
favourite amusement of this little band was putting to death birds
and small animals with little guillotines made for their use."[34]

In Dijon

"The Dijon Commission had 4,000 Vendeans shot. . . . They
were buried in the quarries of Miseri beneath a thin layer of earth,
and the stench of the charnel-house was wafted down to the city,
which it terrified. It was then that a tardy reaction towards pity
took place."[35]

"Robespierre transported the guillotine . . . to an open space
near the Barriere du Trone. There it stood little more than six
busy weeks, in which it dispatched fourteen hundred and three
victims! It was finally conveyed—for Robespierre's own use—to
its original position, in order that he and his friends might die on
the scene of their most remarkable triumphs."[36]

"Robespierre and 21 of his accomplices were sent to the
guillotine on the 28th of July. On the 29th, 70 members of the
Commune were beheaded, and 12 more on the 30th. . . . The
Communist Revolution was dead."[37]

"Thus terminated the Reign of Terror . . . an epoch fraught
with greater political instruction than any of equal duration
which has existed since the beginning of the world. In no former
period had the efforts of the populace so completely triumphed,
or the higher orders been so thoroughly crushed by the lower
. . . ."[38]

---

34 Alison, *History of Europe*, Vol. III, *op. cit.*, p. 101.
35 Mathiez, *The French Revolution*, *op. cit.*, p. 403.
36 Thiers, *History of French Revolution*, Vol. III, *op. cit.*, p. 450.
Footnote quoted from *Quarterly Review*.
37 Gaxotte, *The French Revolution*, *op. cit.*, p. 347.
38 Alison, *History of Europe*, *Vol. III*, *op. cit.*, pp. 354-6.

". . . the nobles were in exile, the clergy in captivity, the gentry in affliction. A merciless sword had . . . [destroyed] alike the dignity of rank, the splendour of talent, and graces of beauty. All that excelled the labouring classes in situation, fortune, or acquirement, had been removed; they had triumphed over their oppressors, seized their possessions, and risen into their station. And what was the consequence? The establishment of a more cruel and revolting tyranny than any which mankind had yet witnessed; the destruction of all the charities and enjoyments of life; the dreadful spectacle of streams of blood flowing through every part of France. . . ."[39]

"The earliest friends, the warmest advocates, the firmest supporters of the populace, were swept off indiscriminately with their bitterest enemies; in the unequal struggle, virtue and philanthropy sunk under ambition and violence. Such are the results of unchaining the passions of the multitude. . . ."[40]

The population of France at this period was approximately 25 million. The historian Prudhomme estimates that the total victims exceeded one million persons.[41]

"The facility with which a faction, composed of a few of the most audacious and reckless of the nation, triumphed over the immense majority of all the holders of property in the kingdom, and led them forth like victims to the sacrifice, is not the least extraordinary or memorable fact of that eventful period. The active part of the bloody faction at Paris never exceeded a few thousand men; their talents were by no means of the highest order . . . yet they trampled under foot all the influential classes, ruled mighty armies with absolute sway, kept 200,000 of their fellow-citizens in captivity, and daily led out several . . . thousand persons, of the best blood in France, to execution. Such is the effect of the unity of action which atrocious wickedness produces; such the consequence of rousing the cupidity of the lower orders; such the ascendency which, in periods of anarchy, is acquired by the most savage and lawless of the people."[42]

---

39 *Ibid.*
40 *Ibid.*
41 Thiers, *History of the French Revolution*, Vol. III, *op. cit.*, pp. 474-75.
42 Alison, *History of Europe*, Vol. III, *op. cit.*, 357-8.

"At last I perceive that in revolutions the supreme power finally rests with the most abandoned." —Danton[43] [as he was led to the guillotine].

"How, then, did a faction, whose leaders were so extremely contemptible in point of numbers, obtain the power to rule France with such absolute sway? The answer is simple. It was ... by promoting ... the cupidity and ambition of those to whom fortune had proved adverse. . . . . Their principle was to keep the revolutionary passions of the people constantly awake by the display of fresh objects of desire; to represent all the present misery which the system of innovation had occasioned, as the consequence of the resistance which the holders of property had opposed to its progress; and to dazzle the populace by the prospect of boundless felicity, when the revolutionary equality and spoliation for which they contended was fully established. By this means, they effectually secured, over the greater part of France, the co-operation of the multitude; . . . this system succeeded perfectly, as long as the victims of spoliation were the higher orders and considerable holders of property; it was when they were exhausted, and the edge of the guillotine began to descend upon the shopkeepers and the more opulent of the labouring-classes, that the *general* reaction took place which overturned the Reign of Terror."[44]

"The ruling principle . . . was to destroy the whole aristocracy both of rank and talent . . . the mass of the people ardently supported a government which was rapidly destroying everything which was above them in station or superior in ability."[45]

"All circumstances taken together, the French Revolution is the most astonishing that has hitherto happened in the world. . . . Other revolutions have been conducted by persons who, whilst they attempted or affected changes in the commonwealth, sanctified their ambition by advancing the dignity of the people whose peace they troubled. . . . Such was . . . Cromwell. . . . Such were your whole race of Guises, Condes and Colignis. . . . These

---

43 *Ibid.*, p. 234. Also in Thiers, *History of the French Revolution*, Vol. III, *op. cit.*, p. 349 footnote.
44 *Ibid.*, p. 360.
45 Alison, *History of Europe*, Vol. III, *op. cit.*, p. 300.

men among all their massacres, had not slain *the mind* in their country."[46]

THE NATURE OF MASS UPRISINGS

In the French Revolution only a million or so people were killed.[47] It was but a small-scale preview of the enormous class revolutions of our own time.

"The first half of the 20th century has seen mankind scale astounding heights of rationality and fall to incredible depths of irrationality. If we fix our eyes on scientific achievement, we discover a world transformed for human use and happiness, but if we think of political history, we peer into mires of passion, hate and murder."[48]

In our century the masses and their leaders, in the process of establishing "dictatorship of the proletariat" over enormous areas of the earth, have massacred their own countrymen on a scale so vast as to exceed in deadliness all other human slaughterings which ever occurred.

". . . communism . . . has destroyed . . . more human beings than any force since the dawn of time."[49]

Highly organized warfare between cities began in Sumer, and has since developed into warfare between nations and between groups of allied nations. Warfare between classes is a later phenomenon. It became super-conflict only after natural selection had been so weakened that those with mediocre mentality could survive and multiply, generation after generation, until they reached such overwhelming preponderance that they could war upon the classes above them.

Of all the phenomena in history, the newer warfare—internal class war—is now the most tremendous. And, by its very nature, it has an enormous influence on the mind and characteristics of man.

------

46 Edmund Burke, *Reflections on the Revolution in France* (New York: Liberal Arts Press, 1955), pp. 54-5.

47 See footnote 41.

48 Robert W. White, *Scientific American*, Vol. 201, Sept., 1959, p. 267.

49 U.S. House of Representatives, 86th Congress, Committee on Un-American Activities, Jan. 13, 1960, in *Lest We Forget!* (Washington, D.C.: U.S. Government Printing Office), p. 1.

2. Class Warfare in Russia

"Marxism . . . rejects the concept of the united people in favor of class struggle."[50]

"The October (1917) Socialist Revolution is of the greatest significance. . . . The Russian proletariat together with the poorest peasantry, under the leadership of the Bolshevik Party headed by Lenin, took power. . . ."[51]

"The Soviet regime . . . is the first in the world (or strictly speaking, the second, because the Commune of Paris attempted to do the same thing) to attract the masses. . . ."[52]

"We come forward as the representatives of 70% of the population of the earth."[53]

"Communist leaders well know that their voluntary followers must be recruited from people who are mentally immature—people who, even though of adult age, have never really grown up."[54]

"The ignorant primarily, but also the vicious, criminal element, which in Russia has murdered all who stood in their way and has robbed all who had any wealth, have accepted the doctrines of communism. No right-minded person can countenance such revolutionary propaganda as the communists are spreading.

"Propaganda has become the most powerful single weapon in the communist arsenal as the means of arousing the masses, luring them toward communism, and preparing and organizing them for revolutionary activity."[55]

". . . witless heads are only too ready to accept anything that fans prevailing discontent, or seems to justify the proscrip-

---

50 Benjamin Schwartz, "The Intelligentsia in Communist China," *Daedalus*, Summer, 1960, p. 608.

51 Nikita Khrushchev, 1957, see *Soviet World Outlook*, U.S. Dept. of State Publication 6836, 1959, p. 88.

52 V. I. Lenin, "The Proletarian Revolution and Kautsky," *Handbook of Marxism*, 1919, p. 828.

53 V. I. Lenin, 1920, see *Soviet World Outlook*, op. cit., p. 138.

54 Fred G. Clark and Richard S. Rimanoczy, *Why Communists Hate* (New York: American Economic Foundation).

55 J. Edgar Hoover, *On Communism* (New York: Random House, 1969), pp. 78, 114.

tion of those who by their skill and energy and industry have enriched society."[56]

"It (Communism) provides a retreat for those who cannot bear the responsibility of individual decision or action. In the party, all decisions are made by the party leaders for the individual. At the other extreme, the party also furnishes those possessed by the will for power the opportunity to dominate and exert authority over others."[57]

"Contempt for death must spread among the masses and thus secure victory . . . the ruthless extermination of the enemy will be their task. . . ."[58]

"The substance of our . . . activity and our program: The overthrow of the bourgeoisie[59] by force, the establishment of the dictatorship of the proletariat, a merciless class struggle on an international scale. . . ."[60]

"Dictatorship is power based directly upon force and unrestricted by any laws. The revolutionary dictatorship of the proletariat is power won and maintained by the violence of the proletariat. . . ."[61]

"The reactionaries . . . speak of the dictatorship as something frightful. They assert that the dictatorship of the proletariat is a cruel power. This is true, . . ."[62]

------

56 Benjamin S. Terry, *The American Journal of Sociology*, Vol. II, 1896.

57 J. Edgar Hoover, *A Study of Communism* (New York: Holt, 1962), p. 9.

58 V. I. Lenin, quoted by J. Edgar Hoover, *Masters of Deceit* (New York: Holt, 1958), p. 35.

59 "Bourgeois means an owner of property. The bourgeoisie are all the owners of property taken together." Nicolai Lenin, *To the Rural Poor*, 1903. See *Selected Works*, Vol. 11 (London: Lawrence and Wishart, 1936), p. 254.

60 "Trade Unions Unite Under the Banner of the Communist International," *Petrograd Pravda*, Aug. 5, 1920: *The II Congress of the Comintern as Reported in Official Newspapers of the Soviet Union* (Washington, D.C.: Government Printing Office, 1920), p. 109. See also *Soviet World Outlook*, U.S. Dept. of State Publication 6836, 1959, p. 118.

61 V. I. Lenin, "The Proletarian Revolution and Renegade Kautsky" (1918), *Selected Works*, Vol. VII (New York: International Publishers, 1943), p. 123.

62 N. Khrushchev in speech at the Csepel Iron Works, Hungary; Budapest radio broadcast, April 9, 1958. *Soviet World Outlook, op. cit.*, p. 15.

". . . 'we kill whole classes,' cried Grigory Zinoviev, party head of Petrograd."[63]

"We are not warring against individual bourgeois. We are out to destroy the bourgeoisie as a class. Hence, whenever a bourgeois is under examination the first step should be, not to endeavour to discover . . . proof that the accused has opposed the Soviet Government, whether verbally or actually, but to put to the witness the three questions: 'To what class does the accused belong?' 'What is his origin?' and 'Describe his upbringing, education, and profession.' Solely in accordance with the answers to these three questions should his fate be decided. For this is what 'Red Terror' means, . . ."[64]

"Bolshevism's cardinal tenets—the dictatorship of the proletariat, and the destruction of the 'classes' by social war—are of truly hideous import. The 'classes,' as conceived by Bolshevism, are very numerous. They comprise not merely the 'idle rich,' but also the whole of the upper and middle social strata, the landowing country folk, the skilled working men; in short, all except those who work with their untutored hands, plus the elect few who philosophize for those who work with their untutored hands."[65]

". . . the Russians had liquidated the middle class and had banned all freedom of private enterprise, not only for capitalists but also for trades-unions."[66]

"The true story of that period will probably never be told in detail. For except for those who escaped to other lands, and they were relatively few, practically the whole upper and middle classes of Russia have been completely exterminated."[67]

"At the same time the dissolution of the old social order was proceeding rapidly. The higher classes disappeared one after another, and by the end of 1920 there were no more landlords, in-

---

63 W. G. Miller, H. L. Roberts, and M. D. Shulman, *The Meaning of Communism* (Morristown, N. J.: Silver Burdette, 1963), p. 71.

64 Latzis, quoted by Sergey Petrovich Melgounov, *The Red Terror in Russia* (London: J. M. Dent, 1926), pp. 39-40, 138.

65 Theodore Stoddard, *The Rising Tide*—(New York: Scribner's, 1922), pp. 218-19.

66 Toynbee, *A Study of History*, Vol. II, *op. cit.*, p. 339.

67 Wendell L. Willkie, *One World* (New York: Simon & Schuster, 1943), pp. 52-3.

dustrialists, private bankers, or big business men in Russia. After another ten or fifteen years most of the small traders and individual [farm-owning] peasants had likewise disappeared."[68]

Communist Masses Liquidate the Intelligent

"The dictatorship of the proletariat is the rule of one class, which takes into its hands the whole apparatus of the new state, which vanquishes the bourgeoisie and neutralizes the whole of the petty-bourgeoisie, the (landed) peasantry, the lower middle class and the intelligentsia."[69]

"Embittered and hardened in exile, or crushed spiritually and physically under the present government, the tragedy of the Russian Intelligentsia is the most pathetic and poignant in human history."[70]

In 1923 Professor Ivan Pavlov wrote to Stalin: "You are depraving and annihilating the intelligentsia to such an extent that I am ashamed to be called a Russian."[71]

"Andrei Vyshinsky . . . was . . . deeply involved in the purge of the academic world."[72]

". . . Stalin's . . . brutal crimes against the 'class enemy'—the (farm-owning) peasantry and the intelligentsia. . . ."[73]

". . . Bolshevism, descending in these last years on Russia, has nearly wiped out her finest people and her best intellect."[74]

"Under Communism, the war against intelligence assumes the form of persecution and physical destruction of the upper classes and the intelligentsia."[75]

---

68 David J. Dallin, *The Real Soviet Russia* (New Haven: Yale University Press, 1947), p. 139.

69 V. I. Lenin, "Tasks of the Third International" (July 14, 1919), *Selected Works*, Vol. X, *op. cit.*, pp. 51-2. See also *Soviet World Outlook, op. cit.*, p. 56.

70 Leo Pasvolsky, "The Intelligentsia Under the Soviets," *Atlantic Monthly*, Vol. 126, November, 1920, p. 692.

71 W. Horsley Gantt, "Pavlov, Champion of the Truth," *Modern Medicine*, November 12, 1962, p. 338.

72 Robert Conquest, *The Great Terror* (London: Macmillan, 1968), p. 17.

73 Milovan Djilas, *Conversations With Stalin* (New York: Harcourt, 1962), p. 188.

74 J. H. Curle, *Our Testing Time* (New York: Doran, 1926), p. 80.

75 Weyl and Possony, *The Geography of Intellect, op. cit.*, p. 145.

"Thousands of scholars, writers, teachers, state officials, even students, were arrested, deported or liquidated in jail.

"Mass repression in the Caucasus in the 1930's led to the exile to Siberia or execution of approximately 422,000 people, drawn particularly from the intelligentsia. . . ."[76]

"The State industrial enterprises, the municipal councils, the educational and scientific bodies—all lost their leaders by the hundred.

"The Russian intelligentsia had for over a century been the traditional repository of the ideas of resistance to despotism and above all to thought control. It was natural that the Purge struck at it with particular force."[77]

". . . the process of conquest by . . . internal proletariats . . . involves the extermination of a large part of the urban upper class and hence of the intellectual elite."[78]

"The deep ideological and political antagonism which, in the early period of the Soviet regime, divided Bolshevism and all the propertied and middle classes of old Russia applied particularly to the intelligentsia. . . ."[79]

"The Soviet policy . . . liquidates most of the creative minority. . . ."[80]

Communists slay or imprison those who are capable of independent thought, even if they have few material possessions, so as to silence possible opposition.

"Intelligents are judged 'enemies of the people' and are purged or liquidated."[81]

"The fate of the middle classes was shared by other elements of Russian society; by the nobility, gentry, capitalists, and 'intellectuals'. The tragedy of the intellectuals is a peculiarly poignant one. The Russian intellectuals, or *Intelligentsia*, as they called themselves, had for generations been Russia's brain and conscience. In the Intelligentsia were concentrated Russia's best

---

76 *Intelligence Digest*, Vol. 24, No. 279, February, 1962, pp. 14-15.

77 Conquest, *The Great Terror, op. cit.*, pp. 257, 317.

78 Weyl and Possony, *The Geography of Intellect, op. cit.*, p. 85.

79 Dallin, *The Real Soviet Russia, op. cit.*, p. 139.

80 Weyl and Possony, *The Geography of Intellect, op. cit.*, p. xi. The number of Nobel scientists per 100 million of population is 175 in Switzerland, 5 in U.S.S.R., *op. cit.*, p. 142.

81 Dr. Ellis A. Johnson (who directed the "Johns Hopkins Report on American Defense," 1956-58), *U.S. News*, Jan. 31, 1958, p. 50.

hopes of progress and civilization. The Intelligentsia stood bravely between despotic Czardom and benighted masses, striving to liberalize the one and to enlighten the other, accepting persecution and misunderstanding as part of its noble task . . . the Intelligentsia was not of one mind. It had its conservatives, its liberals, its radicals, even its violent extremists—from which the brains of Nihilism and Bolshevism were drawn. . . . The intelligentsia backed the *political* revolutions of 1905 and March, 1917. The latter, in particular, fired it with boundless hopes. The Intelligentsia believed that its labors and trials were at last to be rewarded; that Russia was to become the liberal, progressive nation of its dreams.

"Then came the Bolshevik coup . . . [the October revolution, 1917]. The extremist wing of the Intelligentsia accepted Bolshevism with delirium, but the majority rejected it with horror. Bolshevism's narrow class consciousness, savage temper, fierce destructiveness, and hatred of intellect appalled and disgusted the Intelligentsia's liberal idealism. But the Bolsheviks, on their side, had long hated and despised the intellectuals, regarding them as enemies to be swept ruthlessly from their path. The result was a persecution of the intellectuals as implacable as the persecution of the bourgeoisie. The Russian intellectuals were killed, starved, and driven into exile. Multitudes perished, while the survivors were utterly broken. . . ."[82]

Religion and the Liquidation of the Clergy

Certain categories of the intelligent are eliminated with especial rigor. Among these are churchmen.

"Since 1917 Russia has made war on the church and on religion."[83]

Marx had referred to religion as "the opium of the people."[84] Expanding this theme, Lenin had written: "Religion is the opium of the people. Religion is a kind of spiritual vodka in which the

---

82 Stoddard, *The Revolt Against Civilization, op. cit.*, pp. 196-7.
83 Professor F. A. Golder, *The Lessons of the Great War and the Russian Revolution:* (Hoover Library, 1924), p. 5.
84 *Communism in Action*, p. 126. House Document No. 754, 79th Congress, 2nd Session (Washington, D.C.: U.S. Government Printing Office, 1946).

slaves of capital drown their human shape and their claims to any decent human life."[85] In 1909 he wrote: "We must combat religion—that is the rudiment of all materialism and consequently of Marxism."[86] In 1918, having come into power, he ordered the Cheka to "put into effect a merciless terror against the kulaks, priests and White Guards."[87] In 1920 he wrote: ". . . it is necessary to fight against the clergy and other influential reactionary and medieval elements. . . ."[88]

An intensified campaign of assault on religion began in January, 1928. "During the first Communist decade, a large proportion of the clergy of all faiths were wiped out along with other 'enemies of the regime.' Churches were confiscated,[89] schools taken over, the religious press shut down and all religious teaching of the young prohibited.

"Now as the war against the [landowning] peasants gained momentum, the attacks on the Church multiplied. The persecution of all religions in Russia and later in the satellites constitutes a catalog of thousands of crimes."[90]

"Have we suppressed the reactionary clergy? Yes, we have. The unfortunate thing is that it has not been completely liquidated . . . the complete liquidation of the reactionary clergy must be brought about."[91]

Religion was hard to exterminate. In 1937 a third wave of religious persecution struck.[92]

"I spoke a few moments ago respecting the Communist torture of the clergy in Lithuania and said that most of the photographs which had been taken of the corpses of the clergymen

85 V. I. Lenin, "Socialism and Religion," *Selected Works*, Vol. XI (New York: International Publishers, 1943), p. 658. Also: *Soviet World Outlook, op. cit.*, p. 77.

86 V. I. Lenin, "Attitude of Workers' Party," *ibid.*, p. 666.

87 *Facts on Communism*, Vol. II, House Document No. 139 (Washington, D.C.: U.S. Government Printing Office, 1961), p. 87.

88 V. I. Lenin, "Preliminary Theses . . .," *Selected Works, op. cit.*, p. 236. And *Soviet World Outlook, op. cit.*, p. 72.

89 "The state confiscated all church buildings and property." *Communism in Action, op. cit.*, p. 217.

90 *Ibid.*

91 J. Stalin, *Leninism*, Vol. II (New York: International Publishers, 1933), pp. 69-70. Also *Soviet World Outlook, op. cit.*, p. 78.

92 See Dallin, *The Real Soviet Russia, op. cit.*, p. 70.

showed such mutilation that we cannot positively match up the photograph of the corpse with the photograph of the person before mutilization."[93]

"All religion, all churches are prime objects of Communist hatred and in Communist teachings must be destroyed."[94]

"By order of Lenin the All-Russia Extraordinary Committee for the Suppression of Counterrevolution, generally known as VCHK, was established. At the beginning of 1918 this committee started a reign of terror against the clergy as well as other representatives of the old regime. . . . Many outstanding bishops, priests and other clergymen were murdered. . . . Assassinations were carried out by order of the VCHK or its local agencies, by decisions of the revolutionary tribunals, by orders of local executive committees or commanders of the Red Army, or according to lynch law by soldiers, sailors or party activists who were roused by unbridled propaganda. In several towns and rural areas entire groups of . . . clergy were shot."[95]

"At present (1962) in our country (Russia) the activities for the elimination of the religious remnants are going on with increasing speed. . . ."[96]

"The persecution (1964) is no longer bloody, . . . priests and the faithful are imprisoned or shut up in psychiatric asylums . . . according to reliable information, 16,000 priests in the USSR have disappeared. In three years, 10,500 churches have been closed."[97]

". . . the rolls of the clergy had fallen from 50,000 in Imperial Russia to 5,600."[98]

------

93 *Lest We Forget!* A Pictorial Summary of Communism in Action, Consultation with Mr. Klaus Samuli Gunnar Romppanen, Committee on Un-American Activities—86th Congress (Washington, D.C.: U.S. Government Printing Office, 1960), p. 32.

94 *Ibid.*, p. 39.

95 *The Church and State Under Communism*, Vol. I, Pt. 2 and 3, p. 27, Law Library, Library of Congress. Committee on the Judiciary, United States Senate, 89th Congress, 1st Session, 1965.

96 *Ibid.*, p. 35.

97 *Ibid.*, p. 42.

98 *Ibid.*, p. 44. Quoting from Constantin de Grunwall, *The Churches and the Soviet Union* (New York: Macmillan, 1962), p. 57.

"Priests among the prisoners were especially persecuted."[99]

Another category of the intelligent which was intensively killed off was the military officers:

". . . In June ('37) reverberated the thunderbolt which decapitated the General Staff and struck terror into the country: under the unheard-of charge of espionage, under the ridiculous pretext of having 'violated their military oath, betrayed their country, betrayed the peoples of the U.S.S.R., betrayed the Red Army,' Marshal Tukhachevsky, Generals Yakir, Kork, Uborevich, Eideman, Feldman, Primakov and Putna, all well-known 'heroes of the Civil War,' all several times decorated with the order of the Red Flag, all classed as adversaries of Trotsky and partisans of Stalin, were tried in camera, condemned to death without witnesses or defense, and executed within forty-eight hours."[100]

"The military purge lasted until the following year. When it finally was over, the majority of the Soviet marshals, generals, and colonels, and approximately 30,000 officers of lower rank had been killed. Stalin, in his obsession for total power, had crippled Soviet military leadership just one year before the outbreak of World War II."[101]

". . . 90% of all generals, 80% of the colonels, and 30,000 other officers were killed. . . ."[102]

Shifts in Soviet Policy Toward the Intelligent

The initial Russian communist practice, carried out for many years after they seized power in 1917, was to slaughter all classes but the propertyless portion of the laboring class and their communist leaders. This led to the flight abroad of millions of the intelligent in the early years before the Iron Curtain became so effective.

----

99 *Soviet Justice*, "Showplace" prisons vs. real slave labor camps. Consultation with Mr. Adam Joseph Galinski, Committee on Un-American Activities, House of Representatives, 86th Congress, 2nd Session, 1960, p. 40.

100 Boris Souvarine, *Stalin* (Toronto: Longmans, 1939), p. 269.

101 Hoover, *A Study of Communism, op. cit.*, p. 106.

102 Richard M. Ketchum, ed., *What Is Communism?* (New York: Dutton, 1963), p. 57.

"Such an exodus of the educated and the intelligent as there has been out of Russia no country has ever seen, . . ."[103]

When, of the remaining middle and upper classes, enough millions had been slain to eradicate serious opposition, outright slaughter was abated somewhat. Instead, millions of the remainder were put into slave labor camps "to die usefully."[104] Later, as international competition intensified, the communists realized their need for men of intelligence. By then they had some who were young enough to have been reared under communism, with relatively "safe" minds.[105]

There were intensifications in the war against the intelligentsia, and there were relaxations of the war—especially when the intelligents were needed in foreign wars or in five-year programs—but they remained through it all outside the pale of workers and peasants. They will always remain so, being suffered only if the need for them exceeds the hatred of them.

Occasionally excepted in the war against the intelligent were the scientists.

"Following the revolution in 1917, the Soviet authorities were quick to grasp the significance of the scientific community to the nation and provided extensive funds for resumption of the basic research effort. The scientists again became an elite group. Unfortunately, however, they were not exclusive enough to escape the purge that spread throughout Russia in 1929, and many prominent scholars were exiled or disappeared.[106]

"When the purge subsided during the late 1930's, emphasis was again placed on practical research that could be applied to strengthening the military might, and this effort paid off in the early years of World War II."[107]

------

103 Pasvolsky, "The Intelligentsia Under the Soviets," *op. cit.*, p. 692.

104 A Stalinist expression.

105 "The caste system of Tsarist Russia had not permitted the intelligent to rise to any level for which they could qualify, as currently in France and the United States. Accordingly, there were more genes for intelligence buried in the Russian masses than in the masses of some other countries. These were the chief source of bright young people in Communist Russia." Comment of Hermann J. Muller in typescript of *Man*.

106 James W. Useller, *Ordnance*, Vol. 43, Sept.-Oct., 1958, p. 242.

107 *Ibid.*, p. 243.

"The political vacillations of the regime with respect to the intelligentsia continued with remarkable regularity, but the basic attitude was one of distrust. Neither the restoration of industry in the 'twenties, accomplished with the aid of the technological intelligentsia, nor the expansion of scientific institutes and other institutions—nothing could convert the intelligentsia to active support of Communism. Some intellectuals to be sure eventually found their way into the Communist party, but this fact only increased the regime's distrust of the rest.

"The (Shakhta) trial had all the earmarks of a political demonstration, and was regarded as a signal for a new attack upon the intelligentsia."[108]

"Inside Russia there is a visible return to repression, a creeping restoration of the methods . . . of Joseph Stalin's days.

"The crackdown on writers and intellectuals, begun anew three years ago, is still on the rise."[109]

At other times the Russian communists have been reported to reward their remaining intelligent and competent people more, relatively, than such people are rewarded in our increasingly socialized capitalist system. However, our chief concern here is not socio-economic systems but biological effects.

"It is when we reach the Bolshevik Revolution of October, 1917, that a new chapter in the history of internal dysgenic catastrophe opens. Communist doctrine demands the liquidation of the ruling class in toto and this doctrine was applied with systematic thoroughness. . . . The second phase was to decimate, through the Bolshevik purges of all rival revolutionary parties, that substantial portion of the Russian intelligentsia which had supported the overthrow of czarism. The third phase was the Stalinist slaughter of most of the intellectuals who were Bolsheviks."[110]

"The Communists, operating with total disregard of genetics, have exterminated upper classes for political reasons and used such devices as collectivization and man-made famines to obliterate individuals with more than average initiative and brains.

108 Dallin, *The Real Soviet Russia, op. cit.,* pp. 140-41.
109 *U. S. News and World Report,* Vol. LXVI, No. 7, Feb. 17, 1969, p. 57.
110 Weyl and Possony, *The Geography of Intellect, op. cit.,* p. 147.

"If we consider the matter with cold objectivity, it becomes plain that, to the extent that Soviet power and Communist revolution spread, dysgenic deterioration and catastrophe will radiate outward from the present Eurasian core of Communist rule. This means dementalization, both in the form of brainwashing and in the much more irreparable, genetic sense. The conscious purpose and final upshot of these processes of physical extermination of the elite will be to stabilize and equalize human intelligence at a level so low that Homo sapiens will no longer be capable of functioning in a social order based on individual initiative and personal freedom."[111]

"The extent of this genetic havoc is masked by other processes, operating in the opposite direction. The development of mass education under the Soviets and the creation of enormous employment opportunities for scientists, technologists, executives and professionals of all sorts have developed abilities which lay dormant under the more primitive conditions of czarism.[112] The transformation of any nation from an agricultural to an industrial, and hence from a rural to an urban one, multiplies educational and job opportunities for top and middle-echelon intellectual elites. This transformation is not specifically Russian or Communist. It occurred equally dramatically in Japan."[113]

THE NATURE OF MASS UPRISINGS

3. Class Warfare in China

The information from Communist China is meager and confused by propaganda. However, the pattern of slaughter appears to be typical. Only the numbers exceed all other slaughterings.

"Violent revolution is a universal law of proletarian revolution. To realize the transition to socialism, the proletariat must

---

111 *Ibid.*, p. 149.

112 "It does not follow that these abilities would have continued to lie dormant had czarism not been overthrown, for the rate of industrial advance in Imperial Russia during the last pre-World War I years was at least as great as the rate during the Bolshevik era." *Ibid.*, p. 242.

113 *Ibid.*, pp. 242-43.

wage armed struggle, smash the old state machine and establish the dictatorship of the proletariat."[114]

"In China . . . ruthless Communists . . . have herded hundreds of millions into slave labor, into barracks living, . . . have executed millions of dissidents."[115]

". . . the Communists . . . began the consolidation of their grip on power . . . early in 1951, they showed an unmistakable flash of true color. They massacred two million people to 'consolidate the revolution.'

"The massacres began in the villages when Mao Tse-tung's party cadres convened what they called 'speak bitterness' meetings. The cadres prodded the peasants to denounce their former landlords for old crimes, real and imaginary. The landlords took their abuse kneeling, hands tied behind their back, and then were led away and shot. In the cities, the target was anyone who had not thoroughly knuckled under to Communism. Businessmen and intellectuals naturally were suspect."[116]

"Who are the people the (Red Chinese) government is arresting and killing as counter-revolutionaries?

". . . landlords; the comparatively well-to-do or prominent figures of any community, however popular, and intellectuals—professors, doctors, lawyers, newspapermen—who fail to throw in their lot vociferously with the communists."[117]

"An elder or anyone else respected by the village community is a potential leader of the opposition and as such is liquidated, no matter how innocent the partisans know him to be."[118]

"Those who refused to go along with the Communists were shot; the rest are being systematically 'remolded'. . . . Whatever the figure, it is one of the most appalling records of cold-blooded liquidation in history.

"The principle of party supremacy, the report on China suggests, raises special difficulties at present in Red China because

114 "The Proletarian Revolution . . .," *Peoples Daily* and *Red Flag*, March 31, 1964. "Current Background" series 729, American Consulate General, Hong Kong, p. 22.
115 *U.S. News*, Vol. 16, February 6, 1959, p. 58.
116 USA No. 1, April, 1962, pp. 17-18.
117 *Saturday Evening Post*, 224, June 13, 1951, pp. 166-7.
118 Lin Yutang, "Confucius and Marx," *The World's Great Religions*, Vol I (New York: Time-Life, 1957), p. 101.

of the party's mistrust of intellectuals, which extends to scientists and engineers."[119]

". . . the working class and toiling masses cannot defeat the armed bourgeois and landlords except by the power of the gun; in this sense we can even say that the whole world can be re-molded only with the gun."[120]

"Thus far, the Chinese Communists have perhaps been less thorough than their Russian teachers in killing off the most intelligent elements in their population. Nevertheless, the casualties of Soviet Chinese purges have been estimated at anywhere from 15 to 20 million human beings, including amongst them the leaven, the flower and the most promising creative potential of a great civilization."[121]

"I prefer to take the figures of Bishop Quinton Y. K. Wong, a bishop of the Episcopal Church in China, who finally escaped. . . . His estimate is that 40 million people were liquidated."[122]

Cuba is an even more recent example, the first in the Western hemisphere. Reliable information about it is, so far, even less than from behind the Iron and Bamboo Curtains. What there is indicates that the pattern is similar.

"In Communist Cuba, the same or similar processes are occurring in microcosm. The Intellectuals have, for the most part, been driven into exile, imprisoned or shot. . . ."[123]

The same pattern is apparent in other areas where communists gain power: ". . . 380,000 South Vietnamese . . . have fled their homes and farms rather than submit to the Red embrace—not counting the 850,000 who fled North Vietnam when the Reds took over there. The Vietnamese . . . knows who's taking his rice, who is conscripting and kidnapping his children, who is killing his teachers and hamlet chiefs. Ten *thousand* local leaders have been assassinated within the last two years."[124]

---

119 John Strohm, "How They Hate Us in Red China," *Reader's Digest*, Vol. 74, Jan., 1959, pp. 33-4.

120 Mao Tse-tung, "Problems of War and Strategy," *Chinese Communist World Outlook*, U.S. Dept. of State Publication 7379, 1962, p. 90.

121 Weyl and Possony, *The Geography of Intellect, op. cit.*, p. 147.

122 Daniel A. Poling in *Ideological Fallacies of Communism*, 85th Congress, 1st Session, House of Representatives Communication on Un-American Activities, 1957, p. 24.

123 Weyl and Possony, *The Geography of Intellect, op. cit.*, p. 147.

124 *Research Institute Report*, July 2, 1965, p. 2.

Summary of Chapter

These grisly examples of the deliberate extermination, by the masses and their masters, of the other classes *including most of the intelligent ones of their very own people* reveal a type of treacherous savagery not otherwise seen in man. We abhor cannibalism, but here are animals killing their own kind, not even for food. Here is an unprecedented evolutionary phenomenon in which intelligence—the quality which enabled man to prevail throughout the earth—now tends to lead to the extermination of its possessors instead of to their survival. Here is artificial selection leveling downward to the level of the masses; working almost exactly contrary to natural selection. The slaughter of the "haves" is bloody evidence of the power and the intent of the masses when their preponderance becomes great enough and their revolutionary tendencies are inflamed by rabble-rousers. Unless arrested, these vast human slaughterings of the middle and upper classes, among whom are the largest proportion of the intelligent, mark the beginning of a quick reversion of the whole species back to the brute stage.

CHAPTER II

THEY PLAN IT HERE, TOO

"It remains the number one goal of world Communism—whether of the Chinese or the Russian variety—to destroy the United States as the center of non-Communist power."[1]

". . . The Kremlin, holding down hundreds of millions of people and aiming at the rule of the world. . . ."[2]

"We will bury you," said Nikita Khrushchev.[3]

". . . the united international Communist movement directed at . . . the triumph of communism throughout the world."[4]

"First we will take Eastern Europe, then the masses of Asia. Then we will surround America, the last citadel of capitalism."[5]

"War to the hilt between communism and capitalism is inevitable. Today, of course, we are not strong enough to attack. Our time will come in 20 to 30 years. To win we shall need the element of surprise. The bourgeosie will have to be put to sleep. So we shall begin by launching the most spectacular peace movement on record. There will be electrifying overtures and unheard of concessions. The capitalist countries, stupid and decadent, will rejoice to cooperate in their own destruction. They will leap to another chance to be friends. As soon as their guard is down, we shall smash them. . . . "[6]

"The Communist leaders, for 40 years, have repeatedly

1 *U.S. News and World Report*, Vol. 60, April 11, 1966, p. 70.
2 Sir Winston Churchill (see *U.S. News*, Vol. 58, February 8, 1965), p. 39.
3 Reported in Allen Dulles, *The Craft of Intelligence* (New York: New American Library, 1965), p. 56. Also quoted in *Life*, Vol. 51, Pt. 2, October 20, 1961, p. 109.
4 N. Khrushchev, January 6, 1961. *Combined Reports on Communist Subversion*, U.S. Senate, Government Printing Office, 1965.
5 Lenin.
6 Dimitri Manuilsky, in a lecture at the Lenin School of Political Warfare.

asserted that no peace can come to the world until they have overcome the free nations."[7]

"The Russian conspirators smile, then frown; they zig, then zag, but they never really deviate. Bloody world revolution is their final goal. They regard America as the nation which must be destroyed. Do not doubt it."[8]

"In its campaign of subversion the Kremlin uses a colossal force, unique in history, of 500,000 overt or covert agents and a network of special schools to train tens of thousands of professional revolutionaries; and it spends on this about $3 billion a year."[9]

"The Communists are tough. They are determined to undermine and take over free countries—and eventually us."[10]

"Their (Communist) policy . . . is to annihilate us. That is a fearful policy for us to face."[11]

". . . the openly flaunted intention of international communism to bring peoples everywhere under . . . subjugation . . . ."[12]

"History has proved and will go on proving that people's war is the most effective weapon against U. S. imperialism and its lackeys. All revolutionary people will learn to wage people's war against U. S. imperialism and its lackeys. They will take up arms, learn to fight battles and become skilled in waging people's war.

"All peoples . . . unite! . . . fight for . . . socialism!"[13]

"If we keep our hands off, the course of events will almost certainly be toward the development of a huge, stupid and un-

---

7 Herbert Hoover, address on his 88th birthday. Reported in *U.S. News and World Report*, Vol. 53, August 20, 1962, p. 84.

8 J. Edgar Hoover, address to the American Legion. Excerpted in *American Mercury*, Vol. 86, January, 1958, p. 11.

9 *Intelligence Digest*, Vol. 26, No. 305, April, 1964, p. 8.

10 W. Averell Harriman, *U.S. News and World Report*, Vol. 59, July 12, 1965, p. 72.

11 Ellis A. Johnson, *U.S. News*, Vol. 44, January 31, 1958, p. 54.

12 General Douglas MacArthur, Senate Document No. 95 (Washington, D.C.: U.S. Government Printing Office, 1964).

13 Lin Piao, vice premier and minister of national defense, Communist China. See *Nation's Business*, Vol. 54, January, 1966, pp. 40-1.

wieldy proletariat, and the extermination of practically the whole middle class, . . ."[14]

Today it should be starkly clear that the intelligent and competent peoples remaining in the free countries have left to them only a choice between being almost completely annihilated, soon or later, in country after country, or of becoming more vigorous, numerous and influential.[15] They will become either men of destiny or victims of destiny. Rampant today and already frightful in size is a plot for world domination by the masters of the masses. Fundamental to that plot is the calculated extermination of the upper and middle classes, including the intelligent who oppose the masters of the masses or who might oppose them. One of the basic practices of the communists is the annihilation of the thinking classes. For the intelligent, then, the only remaining alternative to eventual destruction by the masses is a strong resurgence in numbers and influence.[16] Any lingering hope of subsidence into innocuous desuetude long ago became futile. Communists will not long tolerate minds capable of penetrating their propaganda.

To regard the world-wide threat of communism as a mere "conflict of ideologies" between communism and capitalism is to overlook communism's avowed purpose of world conquest. The phrase itself is a deceptive term, used by communists.[17] Com-

---

14 Ellsworth Huntington and Leon F. Whitney, *The Builders of America* (New York: W. Morrow, 1927), p. 314.

15 "Let it be understood that we cannot go outside of this alternative: liberty, inequality, survival of the fittest; not-liberty, equality, survival of the unfittest. The former carries society forward and favors all of its best members; the latter carries society downwards and favors all of its worst members." —William Graham Sumner, *Essays*, Vol. II (New Haven: Yale University Press, 1934), p. 95.

"Freedom and equality are sworn and everlasting enemies, and when one prevails the other dies." —Will and Ariel Durant, *The Lessons of History* (New York: Simon & Schuster, 1968), p. 20.

16 The intelligent could, of course, elect to exterminate the revolutionary proportion of the masses before it is able to exterminate them. But this, as human evolution and the survival of the masses shows, is not in keeping with the general disposition of the intelligent. The intelligent, by and large, reject class war and favor the concept of a united people.

17 "In the world today there is a fierce struggle for two ideologies: the socialist and the bourgeois, and in this struggle there can be no neutrals." —N. Khrushchev, *Kommunist*, No. 12, 1957, p. 26.

munism's battle is with capitalism where it finds capitalism, with feudalism where it encounters feudalism[18] and with the intelligent everywhere whom they cannot dupe. The communists would as soon take over other socialist states as any and find it easier than most. Communism has ideological overtones,[19] but it consists fundamentally, by its own leaders' definition, of world-wide class warfare, with the destruction of all but the laboring classes and some members of the Party as one of its major operations.

Communism is based upon a fundamental biological phenomenon: it exploits the rapid increase of people with mediocre to poor intelligence which occurs because natural selection no longer keeps their numbers down.[20] This is the source of inner confidence which gives to collectivist leaders the expectation of total world conquest. Even when the failures of communism to adequately feed and house its people[21] are fully evident to

---

18 "Communists falsely pose the issue as one between communism and capitalism. In reality the struggle is one between tyranny and freedom. —Hoover, *On Communism, op. cit.,* p. 151.

". . . the fact that communism succeeded first in Russia and that it gains ground most rapidly in the more backward nations of the world should not be permitted to encourage a false sense of security . . . the . . . revolution has gained strength during the last century in nearly all nations of Western Civilization.

"Although the hopes it offers are known by many to be illusory and although its promises are vain . . . communism gains ground and will continue to gain, as the proportion of those incapable of seeing through the illusion increases." —E. C. Harwood, *20th Century Common Sense and the American Crisis of the 1960's* (American Institute for Economic Research, 1960).

19 ". . . the Kremlin leaders care little, if anything, for Communism as an ideology. It is merely a weapon to be used in aid of their totalitarian ambitions." —*Intelligence Digest,* No. 345, August, 1967, p. 2.

20 "Darwin's book, *The Origin of Species,* is very important and serves me as a basis in natural science for the class struggle in history." —Karl Marx, *The Correspondence of Marx and Engels* (New York: International Publishers, 1935), pp. 125-26.

"The revolutionary emergence of more and more peoples into the world arena creates exceptionally favorable conditions for an unprecedented broadening of the sphere of influence of Marxism-Leninism." —Nikita Khrushchev, Jan. 6, 1961, quoted by Dean Rusk, Secretary of State, U.S.A. in address to Press Club, July 10, 1961—*U.S. News and World Report,* Vol. 51, July 24, 1961, p. 69.

21 "It was part of the harsh, hollow truth of the Soviet. The country is insufficient in all respects, in food, in production of its resources, in industry. There simply isn't enough of anything—except political indoc-

thinking persons, communism will continue to make gains as the ratio of unthinking people increases. This explains why communists do not even need full-scale war in order to dominate throughout the world. They need only wait until the intelligent are sufficiently outnumbered in the countries not yet under communist power. And they do not anticipate a long wait because, by and large, the intelligent do not maintain their numbers in adequate proportion to the masses.

The slaughter of the intelligent has occurred, and will occur, wherever the masses outnumber the intelligent sufficiently to create an opportunity for communist revolutions. It may be delayed by reforms, counter-ideologies, largesse, appeasement or other faint hopes—all of which attack the symptoms instead of the cause—but it will only be stopped by an alert, virile and growing body of the intelligent who leaven their nations as yeast does a rising loaf, and who offer firm and competent leadership. *The basic answer to the increase of the masses is a proportionate increase of the intelligent.* Fortunately, they need not match in numbers. A hundred intelligent men are far more effective than a hundred of the masses.

It is not sufficient for the intelligent simply to cling to diminishing areas and to diminishing influence within their areas of

---

trination and power—to go around. The Soviet Union is a land of shortages and of hungers." —John Noble, *I Was a Slave in Russia* (New York: Devin-Adair, 1958), p. 52.

". . . the Soviet Union, with 35.5 million more people than the U.S., is shown to be turning out considerably less than half the goods produced in the U.S.

"Now, in the fifth year of the housing push, . . . the cranes stand idle over still-unfinished apartment blocks in many Soviet cities. . . . Millions of Russian families still share dwellings designed for single families.

". . . the average Soviet city dweller has little over 8 square yards of housing space. That is smaller than the health standards set by the U.S. Bureau of Prisons for American jail cells.

"The U.S., with 35 million fewer people than Russia and no (quantitative) shortage of housing to overcome, builds more than twice as many square feet of housing each year as Russia does." —*U.S. News and World Report*, Vol. 58, March 22, 1965, p. 55.

"There is not a socialist country on the face of the earth today that can even feed its own people except for our aid and trade with them." —*Grass Roots Forum*, Vol. 3-1, San Gabriel, California, May 30, 1969, p. 9.

relative freedom. A system in which each person is free to rise to the level justified by his own ability and energy holds small attraction for those whose ability and energy will not lift them far. It does little good to show people that private ownership and free enterprise constitute the most productive system ever evolved[22] if those people have scant capacity for production. When individuals are incapable of producing much of value, every evidence of wealth-producing ability in others makes them, not eager to go and do likewise, but to seize the existing wealth. The communist slogan: *"from* each according to his ability; *to* each according to his need"* will prevail whenever the needy can overpower the able. When poor producers preponderate sufficiently and are shown by rabble-rousers how the use of their collective power will enable them to take everything from those who have more, they are aroused to kill or imprison the "haves" so as to seize their possessions.

It is futile to point out that seizure only divides the existing wealth, while impairing the creation of it so seriously that the masses soon enter periods of great starvation.[23] The masses are not perceptive enough to grasp this sort of cause and effect or at

------

22 "Between 1947 and 1963 the globe's seven most advanced nations (U.S., Canada, U.K., France, Italy, West Germany, Japan) doubled their industrial production and raised agricultural output by nearly 50 percent —the greatest economic leap forward in history. Today, in the industrialized lands, some half billion people share an unexampled material well-being." —*Fortune*, Vol. 70, August, 1964, p. 2.

". . . capitalism has brought with it progress not merely in production but also in knowledge. . . ." —Albert Einstein, *The World As I See It* (New York: Philosophical Library, 1949), pp. 77-8.

23 "The following statistics show the agricultural decline under militant Communism. . . .

"The total harvest in 1921 was only about 40% of the average yearly harvest in 1909-13 while the area sown had decreased by almost one-half and the yield per dessiatine had decreased by more than one-third." —*Encyclopedia Britannica*, 14th ed., "Russia," p. 738.

"Bolshevik agricultural policies finally led to a famine which took approximately 5,000,000 lives during 1921 and 1922."—Hoover, *Study of Communism, op. cit.*, p. 89.

Russia, before 1914 one of the greatest wheat producing nations in the world, and a large exporter of wheat, had to be given wheat in 1921 and 1922 and to buy wheat from Canada and the United States in 1962, 3, 4 and 7. —Ian Grey, *The First Fifty Years* (New York: Coward-McCann, 1967), p. 188.

least they are not strongly deterred by considerations of this type. Their way is to chop down the tree in order to get at the fruit.

The virtues and potentialities of the enormously productive free-enterprise system need to be widely taught. This is essential. But it is by no means sufficient for the intelligent to point out to each other the advantages and rewards of free enterprise, liberty and equality of opportunity. So long as the intelligent decrease in proportion to the whole while the less intelligent, who do poorly under equality of opportunity, continue strongly to out-breed the rest, the trend toward political exploitation of the more capable people will continue, and will eventuate in revolution. No people can long retain freedom unless a sufficient proportion of them understands the difference between freedom and license and is capable of using freedom effectively.[24]

When the areas still led by the intelligent classes are sufficiently reduced in resources or enfeebled by infiltration or confusion, or by the declining proportion of intelligent ones within them, then "comes the revolution."[25]

————

24 "All too slowly, we are waking up to the inescapable fact that, in a modern nation, second-rate minds can never long remain free minds." —Kermit Lansner, *Second-Rate Brains* (Garden City: Doubleday, 1958), p. 3.

"Men are qualified for civil liberty in exact proportion to their disposition to put moral chains upon their own appetites; in proportion as their love of justice is above their rapacity; in proportion as their soundness and sobriety of understanding is above their vanity and presumption; in proportion as they are more disposed to listen to the counsels of the wise and good in preference to the flattery of knaves. Society cannot exist unless a controlling power upon the will and appetite be placed somewhere, and the less of it there is within, the more there must be without. It is ordained in the eternal constitution of things that men of intemperate minds cannot be free; their passions forge their fetters." —Edmund Burke (1791) in a letter to a member of the French National Assembly, *Science*, Vol. 160, May 10, 1968, p. 602.

"Liberty is a need felt by . . . people whom nature has endowed with nobler minds than the mass of men." —Napoleon Bonaparte. See J. C. Herald, *The Mind of Napoleon* (New York: Columbia University Press, 1955), p. 73.

". . . the man who is below the average in economic ability desires equality; those who are conscious of superior ability desire freedom." —Durant, *The Lessons of History, op. cit.*, p. 20.

25 In addition to the Communist parties in each nation, in the United States the black Muslims and similar organizations also prepare for violent revolution. —See Alfred Balk and Alex Harley "Black Merchants

Now is a crucial time in man's evolution.[26] Will the masses and their masters conquer everywhere and substantially exterminate the intelligent, through whom the upward evolution of man can best take place? Will the turning point be swiftly downward for mankind everywhere?

The surest means of survival against what is already the mightiest conquest in all of history, and the hugest slaughter of the intelligent ever perpetrated, is a vigorous resurgence against the militant enemy within and without.

The step most fundamental to that counter-offensive is the strengthening of the ranks of those *with the will and the wits* to resist communism.[27] This is the great underperceived fact in

---

of Hate," *Saturday Evening Post*, Vol. 236, January 26, 1963, pp. 68-74.

". . . people never seem to expect revolution for themselves, but only for their children. The actual revolution is always a surprise." —Crane Brinton, *The Anatomy of Revolution* (New York: Knopf, 1952), p. 69.

26 "There are already signs that our general level of intelligence is incapable of coping with the problems of organization that already exist. In fact any course we take which encourages or even permits a lessening of intelligence is an invitation to extinction." —Berrill, *Man's Emerging Mind, op. cit.*, p. 234.

"If man is to find his way successfully through the labyrinth of difficulties that confront him in the years ahead, he must, above all, use his intelligence. He can no longer rely upon the unforeseeable fortunate circumstance; future mistakes will have consequences far more dangerous than past ones have been. He must divorce himself from unreasoned slogans and dogma, from the soothsayer, from the person whose selfish interests compel him to draw false conclusions, from the man who fears truth and knowledge, from the man who prefers indoctrination to education . . . it is within the range of his ability to choose what the changes will be, and how the resources at his disposal will be used—or abused— in the common victory—or ignominious surrender—of mankind." — Brown, *The Challenge of Man's Future, op. cit.*, pp. 265-66.

"We stand today at a supreme crisis of evolution. . . . Our decisions in the next few decades will . . . commit us to ever-expanding horizons, or extinction of the last . . . hope. If we shirk voluntary decisions, blind forces will choose dead end for us. . . . In us today, life has the power at last to take its fate into its own hands, or to fail its destiny forever." —Elliott, *The Shape of Intelligence, op. cit.*, p. 253.

27 "The next crucial decade in the cold war must be fought, essentially, with the brains we now have." —Alexander G. Korol, *Soviet Education for Science and Technology* (Cambridge, Massachusetts, 1954).

The second decade must also be fought with the brains already born and growing. The third decade must be fought, in large part, with brains

the remaining nations which have not undergone proletarian revolutions. The numbers of the intelligent must be increased in order to keep down the possibility of revolution and to keep the cold war from turning violent or, if violence does break out, to fight with sufficient effectiveness so as to prevail.

What is the fundamental answer to this crucial predicament? *It is for the intelligent to increase the number of intelligent people in their nations.* Education can spread knowledge and improve the utilization of intelligence, but it cannot create intelligence itself. This can be accomplished only if the intelligent have enough offspring like themselves to increase the number of good minds in the world—as they formerly did. This, like so many things, is more easily said than done, but it is by no means impossible. It can be done legally, morally and naturally, within the framework of our present society.

This is the very crux of the matter. On its accomplishment, or the lack of it, hinges the fate of the intelligent and the evolution of man. Will the more intelligent classes of man continue to diminish relatively, until most of the remainder may be exterminated? Or will they overcome their self-chosen infertility and the enormous, growing threat sufficiently to prevail and even to advance to higher levels of intelligence?

---

not yet born. Whether we have more, or fewer, good brains then than now depends upon what we do, or do not do, to see that more individuals with good brains are brought into being.

Part III

A REVIEW AND A LOOK AHEAD

CHAPTER I

A REVIEW

"If we could first know where we are, and whither
we are tending, we could better judge what to do and
how to do it."

— *A. Lincoln*

Thousands of able men have labored to unearth the evidence
concerning man's evolution. Other thousands have sought to
understand and explain the influences which have resulted in
the unequalled status to which he has attained. Out of all these
labors, still progressing, comes the knowledge to enable us to
chart, more surely than could any who came before us, the
position of man on his course at this living moment. We are in
a tense and stirring time, and a decisive one.[1]

We can perceive today the slow but truly transfiguring
effects of that cruel contest, survival of the more fit. We see how,
for millions of years, it tended to select the best-adapted individuals to parent the next generation. We can trace roughly the
way in which it took a microscopic blob of primitive protoplasm
and slowly evolved it into complex man. We see how man was
shaped by the cruel contest into the most intelligent and, largely
because of this, the dominant creature. He is the possessor of
mental power to such a degree as to make it potentially one of
the mightiest influences ever unleashed in the organic world.
When applied to the improvement of living things his mental
powers exceed even the innovation of bi-sexuality itself as an
accelerator of the evolutionary process.

Yet, we also see that man, in the past thirteen thousand years,
has largely freed himself from the natural struggle which built
high intelligence into him. As a consequence, he has ceased to

----

1 "Most critical observers agree that the human race is currently
facing the most crucial period in its history." —V. B. McKelvey, *Science*,
April 3, 1959, p. 875.

develop further in intelligence. On the whole, he may not even maintain the level of intelligence which he reached under severe natural selection. Because there are now so many of him, he has more brains at work than ever before and, due to the accumulation and dissemination of knowledge and techniques, far more for them to work with. Nevertheless, deteriorative influences— the unwillingness of many of the more intelligent to reproduce adequately, and the willingness of many of the less intelligent to out-reproduce and then to destroy the "haves," including those who have intelligence—affect the average quality of mind more powerfully than ever before.

With typical adolescent enthusiasm, man left his natural surroundings and set off on a series of wild and brilliant adventures. With adolescent energy, he created elaborate civilizations time and again, only to see them collapse, one after the other, usually through some eroding process in their late stages. The fact that the very people who had made them function did not sufficiently replace themselves probably contributed substantially to this erosion. At any rate, the individual sources of the greatness of these human systems were not adequately renewed and the systems, including even the great Roman Empire, fell apart bit by bit. Or, when weakened sufficiently, they were overthrown by more vigorous systems, as were ancient Sumer, Babylon, Egypt and more. *After the decline and fall of twenty-seven consecutive civilizations it is clear that man has not previously learned to maintain himself long in an advanced state of material and social well-being.*

We begin to perceive that, instead of increasing the inherent capacity for coping with the complexities of existence, the average of capability within each civilization has, time and again, declined until the civilization relapsed into barbarism or perished.

Man is the heir of more than a score of civilizations, and is now in the midst of the most pervasive and powerful one which ever existed. Yet he appears to be slightly past the zenith of this civilization. Since 1914 he has entered upon great wars of destruction, both between nations and between classes, which have made this the bloodiest century in the entire period of man. This has destroyed much and if it continues, promises to accelerate future decline. Man in general is relinquishing the touchstone of intelligence which made him lord of the earth and is simultaneously accelerating the relinquishment of his present civilization. He has, in fact, so drastically altered his situation over much of the earth that intelligence has actually come to have extinction potential instead of survival value. The relative dying out of the intelligent is passive in the free nations, violent in communist nations and fairly rapid in both.

Man's ascent to his present level, both biological and cultural, has not been invariably upward, but has been irregular. Each advance has reached certain peaks of development and then fallen back. New advances by new peoples have then built upon the remnants of the old. However, the latest biological attainments have not equalled, qualitatively, the best (Cro-Magnon); the latest cultural attainments have not equalled, qualitatively, the best (Greek),[2] though quantitatively and scientifically the attainments of the present civilization surpass those of the past. Furthermore, man appears, like the lemmings, to be rushing toward a condition of overcrowding which can be relieved only by drastic and probably uncontrolled means. Though his present

---

2 The little country of Attica in Greece was about the size of Rhode Island and contained (principally in Athens) approximately 90,000 free people. [Fewer than live in Glendale, California today.] From A. L. Kroeber and T. T. Waterman, *Source Book in Anthropology* (New York: Harcourt, 1931), p. 93.

"The Greek population never compared in size with that of Persia or Egypt, and one of the astounding achievements of Greece is that so few should have done so much. . . ."

"In this small country in the space of two centuries there appeared such a galaxy of illustrious men as has never been found on the whole earth in any two centuries since that time." —Weyl and Possony, *The Geography of Intellect, op. cit.*, p. 103.

Between about 500 and 300 B.C. these people produced no less than 25 of the greatest men who ever lived. Some of these immortals were: Socrates, Plato, Aristotle, Pericles, Aeschylus, Euripides, Archimedes, Sophocles, Aristophanes, Phidias, Praxiteles, Thucydides, Pythagoras, Demosthenes, Miltiades, Hippocrates and Solon.

From this little city-state came some of the most sublime and most imitated architecture, the finest sculpture, the most enlightened government, some of the greatest literature and most comprehensive philosophies yet produced. For centuries its warriors were victorious against overwhelming forces. Yet: "Greece died because the men who made her glory had . . . passed away and left none of their kin, and therefore none of their kind." —David Starr Jordan, *The Human Harvest* (Boston: Beacon, 1907), p. 76.

momentum carries him to greater power and prevalence than ever before, the destructive forces working upon him from within appear to have put him past the peak of fitness of his species, and have begun the inward processes which can lead to the downward cycling of his kind, as they did the trilobites, the dinosaurs and every other species which formerly was dominant.

We see our productive abilities so enhanced by the scientific and industrial revolution, especially the agricultural aspects of it, that new millions each year can be sustained upon the globe with us. Yet, it is not the people whose minds conceive these technical advances who see their progeny multiplied more than the others, as it was under natural selection. It is now the less intelligent classes which take the chief biological advantage of technical progress and multiply themselves prodigiously. We also see the attainments of medical science keeping more of the heavily defective population alive than ever before. The enormous productivity and advances wrought by the intelligent have made possible the greatest multiplication of mediocrity ever witnessed, and at this period in history the intelligent, being in general more addicted to birth control than ever before,[3] fail to attend to the

---

3 For example, the birth rate of the Anglo-Saxons, Dutch and others who founded and built the United States has been reduced by various techniques of birth control *from 8 per couple,* the rate from 1620 until 1820, *to about 2 per couple* in 1950 (except in Appalachia and in certain other instances).

In the years when the colonies which were to become the United States and Canada were developing, the colonists were by-and-large courageous, spiritual and intelligent people.

". . . the settlers who established themselves on the shores of New England, . . . possessed, in proportion to their number, a greater mass of intelligence than is to be found in any European nation of our time." Tocqueville, *Democracy in America,* Vol. I, *op. cit.,* p. 38.

Also, they had children abundantly. "From the time of the first settlements to about 1820 . . . (American) women had an average of about 8 children." —Wilson H. Grabill, *Fertility of American Women* (New York: Wiley, 1958), p. 380.

"For more than two centuries, from the time of the first permanent settlements in America to the early decades of the nineteenth century, the fertility of the American people ranked among the world's highest." —Donald I. Bogue, *The Population of the U.S.* (Glencoe, Ill.: Free Press, 1959), p. 291.

". . . our people must be at least doubled every twenty years. . . . Thus there are supposed to be now upwards of one million of English

increase of their kind. These influences tend not only to induce a decline of average intelligence in man; they threaten to be fatal to the intelligent ones now living.

Whatever the interplay of causes—and we have only touched on some of them—the fact remains that too many of the highly competent are failing to increase proportionately in the stupendous increase of mankind. Many are not even replacing themselves fully in the nations where they are free to do so. This invites increasing exploitation of the remainder of the competent ones—a tendency obvious in the free countries to those who look about themselves perceptively. It leads beyond that, to their violent slaughter wherever the masses gain sufficient preponderance and power. The opportunity awaited by the communist masters of the masses is that time when a free country has too few able leaders or becomes as a whole too unintelligent to follow the ablest they have (a condition from which we already suffer at times). Then, in the resulting confusion, economic decline and military incompetence, wielders of the power of the masses take over, through revolution from within, or attack from without, or both. Once in control, they proceed to slaughter or imprison those with property, those who do not think as their new masters do and, to be doubly secure, most of those who think for themselves

---

souls in North America (though it is thought scarce 80,000 have been brought overseas). . . ." —Benjamin Franklin, "Observations Concerning the Increase of Mankind, . . ." *Magazine of History*, Extra No. 63, 1755, quoted in Bogue, *op. cit.*, p. 292. (Another source gives 1767 for this publication.)

These highly intelligent people, multiplying energetically, founded the nations which have become the most productive, most powerful and most generous the world has ever known. At a time when, victorious in World War II and the sole possessors of atomic bombs, with which they could have subdued the world, they relinquished all territorial conquests (including great areas of Italy and Germany, and all of Japan) and assisted enemies and allies to rebuild. The total of U.S. foreign aid alone, to date, exceeds 122 billions of dollars.—*U.S. News and World Report*, Vol.LXIV, No. 18, April 29, 1968, p. 1.

Indians had possessed the same region for thousands of years but made little of it. The Spaniards seized the richest and most civilized portions of the new world, but built no such nations, although they too could draw upon every technical resource of European civilization. The combination of intelligence and fertility, which the founders of the United States and Canada exemplified for centuries, are unsurpassable combinations.

at all, for they know that keen minds penetrate their specious propaganda.

This is a grim prospect for the person capable of thinking for himself. It calls for great decisions—the most momentous in our individual lives and in the entire history of man. Probably our Earth is today about halfway through its cycle of existence as a possible abode of life.[4] What will be the fate of man through the five billion years that may well remain before the earth grows too cold? Will the masses be permitted to continue their engulfment of the intelligent? Will man in general retrogress on a worldwide scale into a more stupid creature with more and more handicaps in his constitution? Will high intelligence eventually prove to have been self-extinguishing; a biologic failure as a result of its own interference with natural selection? Or will man use his new knowledge of the ways of life so as to resume his evolutionary advance? This is the most portentous of all human questions. Today is the most supremely critical period in all of man's evolution since that fearful time when his distant forebears descended from the trees and faced the savage predators. Unless great decisions be made and followed through, man's poorer elements will increasingly engulf and extinguish the more intelligent.

----

4 Harrison, *What Man May Be, op. cit.*, p. 262.

CHAPTER II

A LOOK AHEAD

What, specifically, are the alternatives?

Must man revert to pre-agricultural, ice-age natural selection which lifted him so high by being so fatal to the unintelligent? Short of a cataclysmic return to the hunting stage, he could not do so even if he would.

Must he then continue the present drift toward stupidity, so fatal to the intelligent?

He need not. No, he need not.

There is a way of doing which is infinitely more humane than natural selection. It can elevate all of life which is touched by it.

No one man devised it, but many superior minds contributed to it. It is mankind's greatest cumulative discovery. It is only hidden from most eyes because it transforms gradually instead of swiftly. With it man improves for his use innumerable living things about him. With it he may improve not only his environment but his very own self.[1] With it he may increase his inner resources until he is not only equal to the problems which press upon him but surmounts them to rise to greater heights than any he has yet known. One of the several names for this discovery, the logical successor to cruel and enfeebled natural selection, is intelligent selection.

Man is young as species go. He need not accept senscence

------

1 "What, then, are the prospects of evolving higher intelligence . . . ? Evolution of greater brains is, of course, a practical possibility; and it will be all to the good. . . . Individual excellence can hardly evolve too far." —Elliott, *The Shape of Intelligence, op. cit.*, p. 247.

"Men can be upgraded. At the strictly biological level there is no question that the principles of genetics and selection, used widely by man in the plant and animal world, could be used to enhance desired traits in man." —R. W. Girard, "Intelligence, Information and Education," *Science*, Vol. 148, May 7, 1965, p. 763.

and decline as inevitable. Already he has within his grasp the knowledge to make of himself in time not a poorer but a better man than his forebears. In intelligent selection he has been given the opportunity to control his own evolution. He may scorn to use the knowledge, even as the mighty Cro-Magnon hunters appear to have spurned the grubby techniques of stone-polishing and agriculture and so vanished because, though superior to their successors, they were not equal to the changing situation. Or man may utilize the constructive evolutionary powers which some of his greatest minds have placed ready to his hand. Whether he uses or spurns his new powers will determine whether his future shall be upward or downward.[2] It will be one or the other because change, sometimes swift, sometimes slow, is the first law of the high organic world of which he is a part.

For more than ten thousand years man has had cultural progress at the price of genetic stagnation or regression. It is not

––––––––

2 ". . . dazzling prospects are not guaranteed. Man's upward evolution appears to have hit a snag. Not only has it stopped, but it appears to have gone in reverse. As we ponder the charts, we are reminded of the tunicates, perhaps, and wonder whether man may have met an obstacle insurmountable for the species, though the solution is clear to that small percentage of the human race willing and eager to see. And so we sense two essentially different paths leading to the future.

"Along one . . . we are now idly strolling, scarcely noting its easy decline. The other path does not yet exist. We must build it as we go, step by step, and the heights which we may thus scale are proportional to the wisdom with which we build. The first few steps have been constructed or designed by an intellectual minority. Time only will tell whether the mass of humankind will turn to assist in the task, or whether in blind and uncomprehending rage, they will tear down what appears to them only a barrier to their comfortable journey downward to extinction." —Pauli, *World Of Life, op. cit.*, p. 601.

"There appears to be a finite possibility that, given adequate research and broad planning, deterioration of the species might eventually be halted." —Brown, *The Challenge of Man's Future, op. cit.*, p. 263.

"Most of man's misfortunes are due to his organic and mental constitution and, in a large measure, to his heredity." —Alexis Carrell, *Man the Unknown* (New York: Harper, 1939), p. 300.

"For the first time in the history of humanity a crumbling civilization is capable of discerning the causes of its decay. For the first time it has at its disposal the gigantic strength of science. Will we utilize this knowledge, this power? It is our only hope of escaping the fate common to all great civilizations of the past. Our destiny is in our hands. On the new road, we must now go forward." —*Ibid.*, pp. 321-22.

necessary for him to pay such a price. He may have continuing progress in both areas simultaneously. Each may reinforce and accelerate the other to a degree not heretofore known to man. A new chapter in his evolution can begin and it can be more wonderful than any yet known.

Intelligent selection is no new and untried way of doing. Since man first domesticated plants and animals, he has used it with increasing sureness and effectiveness. With it he has taken the sour wild crab apple and transformed it into many sweet and delicious forms, all with several times more edible substance upon them. With it he has taken wild chickens and evolved therefrom birds more than double in size, and others with many times the egg-laying ability of their ancestors. With it he has taken wild dogs and created animals of greater size, others with greater speed, and still others with quicker minds. In intelligent selection man has an opportunity never before presented to any species on earth. He is in a position to transcend the limitations of the natural selection which for so long set his course. He has a new resource for dealing with his destiny.

Intelligent selection is a way of doing whereby the more intelligent the people, the more of their kind they will bring into the world. Instead of bringing into being the less intelligent and then killing them off at an early age, as natural selection did, it would simply retard—voluntarily for the most part[3]—the rate at which the unintelligent breed, and at the same time encourage the intelligent to bring into being many more like themselves. It is elementary procedure for any improver of animals or plants, yet with ourselves it is still impeded by emotions and attitudes which are left over from our distant and more ignorant past. This is the reason it remains the one great step within the present capabilities of man which he has not yet taken.

Man has imposed intelligent selection upon the forms of life domesticated by him but has not yet seriously applied it to himself and no one has imposed it upon him. Yet intelligent selection need not be imposed. Like government with the consent of the governed, it may be employed voluntarily by any people con-

---

3 "New techniques for inducing temporary infertility . . . . hold much promise." —*Science News Letter*, January 24, 1959, p. 54.

cerned enough and enlightened enough to put it into effect.[4] Man is self-domesticated. He may also be self-selecting. It is as though a beneficent Providence, seeing man faltering and beginning to retrogress, chose to reveal to him in his time of need certain of the secrets of his being, that he might lift himself from the self-impairment and degradation brought on by his own indiscriminate breeding.

Despite his self-imposed impediments, with such a resource at his command, should man permit himself to deteriorate when he could increase his powers? Should the intelligent everywhere remain passive until the masses engulf them? Only lack of realization of the situation, or downright inertia, prevent man from benefiting inherently from his own greatest discovery—prevent him from following farther the path toward higher intelligence upon which nature once set his feet.

We have learned how man has been able to control his environment to such an extent that it no longer continued to improve him. The steps in this control: weapons, tools, fire, agriculture, industries and science—great though they be—pale in importance before the potentialities of the step now open before him. The resources of intelligent selection, capably utilized, can enable him to rise above the limitations, not only of himself and his environment, but of the evolutionary process itself. He may break the cycle of rise and regression which has been characteristic of all dominant organisms and of his own civilizations. He may ascend toward a new level of being, of which his present animal nature is only the crude beginning. He may become, not just the latest creature in a succession of dominant creatures, but the first to direct and shape his own future nature. The mastery of fire was man's first conquest of the forces of nature; the mastery of his own evolution will be his greatest.

"When . . . we consider the logical consequences of the application of the science of genetics to man, we see progress of a hitherto inconceivable kind opening out before us."[5]

----

4 For example, the avoidance of inbreeding and incest, which man has so widely and successfully accomplished, may be regarded as great eugenic attainments, achieved in most part voluntarily and enforced not so much through laws as through social attitudes.

5 Herman J. Muller, *Out of the Night* (New York: Vanguard, 1935), p. 78.

"A vast but unused power is vested in each generation over the very natures of their successors—that is, over their inborn faculties and dispositions."[6]

"Man . . . alone can hope to transcend the blind processes of mutation and natural selection in guiding his future evolution. This ability is itself perhaps the grandest achievement of all organic evolution, for in a sense it gives man the potential power to be the complete master of his own evolutionary destiny."[7]

Intelligent selection can work much more rapidly and benignly than natural selection ever could.[8] It offers the greatest choice ever knowingly made by man.

"Man in this moment of his history has emerged in greater supremacy over the forces of nature than has ever been dreamed of before. He has it in his power to solve quite easily the problems of material existence. He has conquered the wild beasts, and he has even conquered the insects and microbes. There lies before him, as he wishes, a golden age of peace and progress. All is in his hand. He has only to conquer his last and worst enemy—himself."[9]

Fortunately the choice lies with the intelligent, if they do not act too late. Great will become the families, the peoples and the religions which pursue intelligent selection. If a major nation were to pursue intelligent selection diligently, and have so much as a one-generation lead over the rest, no other nation could ever again come abreast of it.[10]

---

6 Galton, *Hereditary Genius, op. cit.,* p. XIX.
7 Beadle, *The Physical and Chemical Basis of Inheritance, op. cit.,* p. 11.
8 ". . . the rates of 'evolutionary' change . . . under artificial selection . . . are tremendous compared with even the more rapidly evolving natural forms—perhaps thousands of times as fast." —James F. Crow, *Scientific American,* Vol. 201, No. 3, September, 1959, pp. 143-44.
9 Sir Winston Churchill, *Commons,* March 28, 1950.
10 "I return to the question of the tendency of civilization to eliminate its ablest people. This has happened in the past and is certainly happening now, and if it is always to happen, it signifies a recurrent degeneration of all civilizations. If any civilized country could overcome this effect, so that it alone retained both its ability and its civilization, it would certainly become the leading nation of the world." —Darwin, *The Next Million Years, op. cit.,* p. 152.
In this period of highly technical warfare it becomes all the more evident that the most critical and strategic commodity of all is the brain-

power of a nation. Great nations have fallen through just a small relative deficiency of it. Others have prevailed despite deficiencies in manpower and material. Today the fall will come swiftly and surely if we even lag in the brain-power race.

". . . back of any great new military, industrial, or technological achievement, back of any nation's material strength, lie the minds of many people. The most important resources of any nation are its intellectual resources. New developments come from new ideas. New ideas develop in the minds of men. Any single new idea, in fact, develops in the mind of a single man; the pooled ideas of many men constitute a nation's intellectual resource.

"Our first task, then, is to strengthen our intellectual resources." —Lee A. DuBridge, "The Challenge of Sputnik," *Engineering and Science*, Vol. XXI, February, 1958, p. 5.

". . . if the people of one nation were to apply them (techniques of selection) intelligently and extensively, even a few decades before the rest of the world did so, they would be able soon afterwards to rise to a so much higher level of capability as to make them virtually invincible." —Hermann J. Muller, *The Next Hundred Years*, Seagram Symposium, November 22, 1957, p. 34.

Part IV

THE OPPORTUNITIES OF THE INTELLIGENT

CHAPTER I

REDUCING THE PRODUCTION OF DEFICIENT
MINDS

"Man is gifted with pity and other kindly feelings;
he has also the power of preventing many kinds of suffering. I conceive it to fall well within his province to
replace Natural Selection by other processes that are
more merciful and not less effective."[1]

We may lift our eyes unto lofty goals ahead but the path
toward them is to be trod one step at a time. Nor is ascent as easy
as descent. Is it possible, then, voluntarily and within our laws
and mores, to put intelligent selection to work for the sake of
man? There are indeed many good ways to do so.

The plenary solution to the great problem described in this
book, and the essence of intelligent selection, is for the intelligent
to release much of the natural fertility which they have repressed
so long, and at the same time assist the mentally deficient voluntarily to reduce their output of offspring.[2] Like many things, this

---

1 Sir Francis Galton, *Memories of My Life* (New York: Dutton,
1908), p. 322.

2 "Says Mr. H. G. Wells: 'It seemed to me that to discourage the
multiplication of people below a certain standard, and to encourage the
multiplication of exceptionally superior people, was the only real and
permanent way of mending the ills of the world. I think that still.'"
—Caleb Williams Saleeby, *Parenthood and Race Culture* (New York:
Moffat, Yard & Co., 1909), p. 15.

"The great problem of civilization is to secure a relative increase
of the valuable as compared with the less valuable elements in the population." —Theodore Roosevelt, *Birth Reform*.

". . . there is now no reasonable excuse for refusing to face the fact
that nothing but . . . eugenics . . . can save our civilization from the
fate that has overtaken all previous civilizations." —George Bernard
Shaw, *Sociological Papers* (London: Macmillan & Co., Ltd., 1904), pp.
74-5.

"The hope for the future lies in the sense of responsibility of the
better stocks in the race. If they increase their output of children . . .

is more easily said than done. But it can be done. The sooner and better it is done the better will be the condition of mankind.

This logically brings us to a discussion of measures for reducing the rate of reproduction of that lowest, careless segment of the population which has remained most primitive in its reproductive habits. Among this segment are most of the mentally deficient and defective specimens.[3] Yet the discussion here will be brief for this aspect of the problem is already much attended to by others[4] with increasing, though not yet adequate, results.

Many societies of today do not believe in killing humans except as punishment for the most extreme crimes, and, of course, during wars. That is to say, they do not ordinarily believe in

the nations of the world can reverse the bad selection of the last seventy years, and gradually improve their average of health, beauty and ability." —Dampier, *A History of Science, op. cit.*, p. 332. This passage was written prior to 1947.

". . . there are two possible ways to change people. First, one can change the environments in which people grow up—the nutrition, sanitation, education, political systems, etc. . . . Secondly, one may strive to decrease the number of people born with hereditary defects of all kinds, and to increase the number with genes which favor the development of superior qualities. . . . It should be obvious that (these) are complementary and not conflicting modes of action."   —Dunn and Dobzhansky, *Heredity, Race and Society, op. cit.*, p. 83.

3 It has been pointed out (page 91) that the largest reproducers are the morons. It is specifically of this segment that we speak here.

"Feeble-mindedness is an absolute dead-weight on the race. It is an evil that is unmitigated. The heavy and complicated social burdens and injuries it inflicts on the present generation are without compensation, while the unquestionable fact that in any degree it is highly inheritable renders it a deteriorating poison to the race; it depreciates the quality of a people." Havelock Ellis, *The Task of Social Hygiene* (New York: Houghton Mifflin, 1912), p. 43.

"Moron children are very rare among parents of normal or superior intelligence, and very frequent among parents of low intelligence—more than three-fourths of the morons coming from one-tenth of the families." —Amram Scheinfeld, *The Basic Facts of Human Heredity* (New York: Washington Square Press, 1961), p. 129.

4 Planned Parenthood—World Population makes a very logical approach to this problem by getting contraceptive information and materials to those who need them but have not had sufficient access to them. The work of this organization is basic, sound and increasingly effective. It justifies your generous support. Address:

Planned Parenthood—World Population
515 Madison Avenue
New York City, New York 10022

taking lives once begun. However, they do sanction the prevention of lives before they are begun. This is because the death of
cast-off sperm or egg cells is not the same order of tragedy as the
death of the living human which can develop after a pair of such
cells unite. Women who bear children to their utmost still produce many an ovum which dies infertile.[5] Every man wastes
hundreds of millions of spermatozoa for each one which joins an
ovum to produce a child.[6] We lock up criminals and the insane,
thus impairing their opportunity to reproduce. Nuns and priests
are forbidden the reproductive act. We widely practice various
methods of preventing conception in order to reduce the fre-
quency of reproduction. We even look with complacency upon
childlessness among highly able and intelligent people, although
that represents irretrievable loss of some of the most precious
human genes. Accordingly, to reduce also the reproduction of
deficient and defective individuals calls for no great change in
attitude, especially if this reduction be largely voluntary, as it can
be with simple techniques of sterilization and contraception.[7]

We are now alert to the communicable diseases and have

---

5 "A normal woman will produce about 400 ova during her fertile
years. If her childbearing is unrestricted, an average of 15 of these will
develop into live children." —Alan Guttmacher, M.D.

6 ". . . accurate conclusions can be drawn from a study of the
semen. The ejaculate . . . average volume is from 3 to 4 c.c. with rather
wide variations. . . . The spermatozoa should number 100,000,000 or more
per c.c." —Austin I. Dodson, Synopsis of Genitourinary Diseases, 5th
ed. (St. Louis: Mosley, 1952), p. 185.

7 Voluntary reduction of the offspring of the incompetent would
seem to hinge, in good part, upon methods of birth control adapted to
those least capable of controlling their births. So far, birth control has
been a two-edged sword which mostly has cut the wrong way; control
is in general most exercised by those whom we least want to exercise it.
Research into methods of simple birth control, adapted to the simple-
minded, is important.

". . . couples at the lowest educational level have a contraceptive
failure rate up to 50% greater than those at higher education levels."
—Five Year Report, American Eugenics Society, 1964, p. 6.

"Through a complete practice of birth control, extending to the
lowermost sections of the population, mankind could leave behind for
ever, within the space of this century, the aimless suffering, the vacillating
advance and chronic relapses which have characterized civilizations since
the culture of Sumeria first blossomed." —Cattell, The Fight for Our
National Intelligence, op. cit., p. 131.

gained substantial control over them. To the hereditary defects and deficiences we are still apathetic. Yet, viewpoints and practices can be modified so that they will no longer lead to the further increase of those who are a burden upon their nations. Why, for example, should we continue to finance the breeding of more and more incompetents for ourselves and our children to support?[8]

The ignorant may be taught; the wicked may be reformed or disciplined into less vicious ways; but against innate stupidity, what can be done?[9] Short of eliminating most of the stupid, as natural selection did, we can only strive to reduce stupidity in the oncoming generations. By arresting the multiplication of stupidity, man's most degrading influence can be lessened, or at least be prevented from growing worse.[10] If we were to expend

---

8 Working against the diminution of the deficient are multi-billion dollar programs of relief. Many of these finance the increase of non-self-supporting types. However beneficent and politically expedient public relief may be, as it is presently carried out it constitutes a biologic blunder of tremendous magnitude. The world will be forever the worse for it. Yet, the beneficence could be retained while correcting the blunder.

Let those of any age and degree of incompetence receive largesse, as long as that be the bent of our public compassion, but let additional facts also be taken into account. The reproductive years are also the years of greatest vigor and capability. If, in an economy as favorable as the present one, an individual cannot support himself and family during the prime of life, whenever can he? Let such individuals be prevented from producing additional chronic parasites like themselves while they are on public relief. (Exceptions: Those on relief because of disabling accidents or other non-hereditary causes.) Let them be offered a simple means of birth control and, if they refuse to use it, withdraw their support. Let them go to the poorhouse if they must, as they did before the day when the handout was brought to them. Most would seek work first and thus have less time to breed.

9 The ancient Greeks had a saying: "Against stupidity the Gods themselves fight in vain." Fortunately, contraceptive science is today providing weapons against even this stubborn problem.

10 Some who think they are preserving human rights seek to maintain that, when the accidents of heredity make a person a liability to himself and others, that person should be free to inflict a similar liability on his little children, and so on and on to countless descendants. This is surely one of the curiosities of human misunderstanding. Let our hearts and our help go out in compassion to unfortunates, but if their affliction be hereditary, let them be urged and assisted not to have children when the children would be likely to bear the same affliction. Let them not transmit and multiply their defectiveness. To maintain that they should be free to do so is like arguing that a leper should be free to infect his own children. *(continued on next page)*

as much money and effort in discouraging the multiplication of deficiencies as is now spent in helping the deficient to multiply, an enormous reduction in the numbers of the unfortunate would be accomplished. Furthermore, the tremendous and growing burden of support would rapidly be lifted from the capable, leaving them freer for more constructive pursuits.

". . . the most real and persistent springs of evil are only stupidity and ignorance, in their widest ramifications. The wisest of the last few generations came to that truth and they struck [in education] a giant blow at ignorance. It is for this generation to banish stupidity—defective intelligence."[11]

It is not that unintelligent people should be prevented from breeding. That is compulsion and is justified only in extreme cases such as persistent criminality. They simply should not be encouraged and financed to breed. Further, they should not be induced to breed by withholding from them knowledge and means of how not to breed. A principle exists which justifies all the emphasis that can be commanded: *Let us be as charitable to the unfortunate as our circumstances permit; but let us not assist those with hereditary deficiencies to multiply their misfortunes*, inflicting them on their offspring.[12] If we continue to assist this process, the average of mankind will become increas-

---

In a state of nature, sexual gratification sufficed to get the job of reproducing done. It was a loose and largely haphazard way, but so long as natural selection weeded out the mistakes there was general improvement despite the innumerable individual tragedies. However, with natural selection now enfeebled, sexual gratification among the careless has become one of the main determinants of the character of the next generation. The effect on the nature of man today is what we should expect from a determinant such as this.

11 Cattell, *The Fight for Our National Intelligence, op. cit.*, p. 162.

12 "Anyone who doubts the often poor inheritance of the mentally retarded person should read the big book telling of the magnificent study made by the Doctors Reed, of the University of Minnesota, of seven generations of mentally retarded people. Another fact that every physician should know is that when some genes which govern the development of the brain go wrong, several different conditions can show up, ranging from psychosis to eccentricity, neurosis, alcoholism, minor equivalents of psychosis, criminality, feeble-mindedness, and epilepsy.

"This fact was proved up to the hilt during a study made years ago by Dr. Serge Androp (Eugenics Research Association, Monograph Series No. 10; Cold Spring Harbor, October, 1935). Because Dr. Androp had spent thirteen years in the little city of Gallipolis, Ohio, he knew well

ingly incapable. Simple means, enabling those with transmissible defects more effectively to control their births, are being rapidly developed and should be made more widely available to them. Let them be assisted and encouraged *not* to breed.

---

the members of some 200 families, many members of which had been cursed by a bad nervous system; also, because he had access to state and hospital records of the antecedents of these people, he was able to make and publish 56 charts—astounding charts—showing how several nervous abnormalities had come down through several generations of certain very unfortunate families.

"For instance, in one case (the S family) in which a feebleminded man married an intelligent woman, in two generations, they produced 4 feebleminded children, 1 epileptic, 1 insane person, 3 criminals, 1 alcoholic, 1 suicide, 1 immoral woman, and 1 pauper.

"In the M family (p. 52) there were 4 feebleminded, 2 insane, 4 epileptic, and 5 criminal persons. In the McM family (p. 27) there were 5 feebleminded, 4 epileptic, 4 insane, 5 violent-tempered, and 1 alcoholic person. In the L family (p. 45) there were 9 feebleminded, 1 criminal, and 2 insane persons."   —Walter C. Alvarez, in *Modern Medicine*, March 11, 1968, p. 85.

CHAPTER II

INCREASING THE PRODUCTION OF GOOD MINDS

Complementing the steps to diminish the flood of morons are measures to encourage the increase of intelligent people in the world. These measures, by multiplying the fund of intelligence, can put many more good minds to work on the problems of mankind. Seven of these Opportunities of the Intelligent are described briefly here.

OPPORTUNITY A:

NATURALIZE HIGHER EDUCATION

In our coeducational colleges we have a vast system, hardly in existence a century ago, wherein a selected group of intelligent young men and women may meet. Marriages among them often occur, especially after graduation. Insofar as colleges bring the intelligent together and encourage, at least by propinquity, their marriage, they contribute to the concentration of intelligence in some very fortunate children. But, insofar as the economic dependency of higher education discourages and delays marriage and the having of children among the intelligent, the situation tends to diminish the total fund of human intelligence.[1] The world in general is experiencing an enormous population explosion but there is no population explosion among the highly educated.

"College graduates did not replace themselves."[2]

---

1 This it does to a very great degree. While college graduates today have more children than they once did, this is no cause for complacency. What has been called a rise in their birth rate is merely a slight reduction in birth control. The highest reproductive rate found for any group of U.S. college graduates is 2.39, while it is 3.65 for the general U.S. population and so must be greater than 3.65 for the less-educated. (Bajema, *Eugenics Quarterly, op. cit.,* p. 306; and Lewis Dublin, *Fact Book on Man* (New York: Macmillan, 1963), p. 50.)

2 Pendell, *Sex vs. Civilization, op. cit.,* p. 108.

163

". . . fertility and education are inversely related. . . ."[3]

". . . it appears that . . . birth control is most widely practiced generally among persons of more education. . . ."[4]

Therein lies the irony of our present educational predicament. The more knowledge amassed by our society, the more years must be spent in higher education by those who would learn even one special part of it. The more intelligent the individuals and the fuller the training and use of their intellect, the longer will their education take, and the later in life will they come into position to support a family. Consequently the smaller will be their family, as a rule, and the fewer will be the intelligent individuals whom they contribute to the next generation. Searching out our best minds and training them fully represents efficient use of existing brainpower, but it also constitutes scraping the bottom of the barrel of this commodity while discouraging its replenishment. It is short-sighted exploitation of our most precious natural resource. It is like searching out the remaining specimens of hardwood trees so as to lumber them off, without providing for the reseeding of their variety. It permits a great cultural advance—higher education—to induce genetic regression. This vicious cycle tends to reduce the proportion of highly educable members of society, generation after generation. Continued long and intensively, it can lead to the decline and overthrow of that society.[5]

---

3 Mitra, "Education and Fertility," *Eugenics Quarterly*, Vol. 13, No. 3, September, 1966, p. 219.

4 *U.S. News and World Report*, May 9, 1966, p. 44.

5 ". . . the educational system of the country acts as a sieve to sift out the more intelligent and destroy their posterity. It is a selection which ensures that their like shall not endure." —Sir Godfrey Thomson, *The Trend of National Intelligence* (London: Eugenics Society, 1947), p. 3.

". . . education (is) not as sudden as a massacre, but . . . more deadly in the long run." —Mark Twain (quoted in *Changing Times*, July, 1960, p. 48).

". . . the outlook remains disquieting. The intellectual work of the world, on which depends continued progress, and indeed the maintenance of the general standard of life, is done by a small fraction of the people, drawn for the most part from the classes who have cut down their output of children. . . . Scholarships and other means of advancing the able from all classes may supply the deficiency for a time, but the amount of ability in the country is limited, and is proportionately less in the lower ranks of society. If it be steadily picked out and raised from those

Yet, if the cycle be intelligently changed into an ascending spiral, it can permit that society to flourish beyond all previous experience. There is nothing naturally incompatible between gaining a higher education and simultaneously having children.[6] It is not that students are willingly dependent during this potentially most fertile period of their lives. It is the system of extended education and its consequent extended dependency which makes them less fertile.

It is necessary to realize how essential it is that our more intelligent young men and women should not be led to defer the having of children until their years of greatest fecundity are past. Rather, they especially should be encouraged to have children during the years when they are naturally best suited to do this, even though they are not then self-supporting. This is an expensive conviction to come to. Nevertheless, the adverse economics can be improved, at least for the most deserving couples, by other intelligent individuals who have more income than the students. Those who are past their own reproductive years—and more affluent than students—can contribute to the great opportunity and direct it to a significant degree.

This is a constructive plan, not a destructive one. It is evolutionary and gradual, not revolutionary or violent. But it is nonetheless militant and urgent. It contemplates the day when, on the campuses of our colleges, so many of the finest and brightest young men are fathers; so many of the best young women are married and having children that more and more of the other students will feel reproached by their own unnatural barrenness and will themselves be caught up in the spirit of the next advance of mankind—and marry earlier so as to participate in it more fully.

Our society annually spends hundreds of millions of dollars to support indolent people of low intelligence while they breed. Can it not do at least as much for its more desirable young couples? The bright and healthy children which college couples can produce are far less a burden than the many deficient children

ranks, it will be partially sterilized as it rises by a decreased birth-rate, and will leave behind it the dead level of an unintelligent proletariat. Gradually the strains of ability will be weeded out of the nation, with ever increasing danger to civilization." —Dampier, *A History of Science, op. cit.*, pp. 330-31.

6 The *rearing* of the children is another matter, discussed later.

which society now finances through relief. Furthermore, when mature, the children of college students usually become assets and supports to society instead of liabilities. Expense may deter, but it does not prevent, and it would be difficult to imagine an investment which, in the long run, would yield finer human returns.

Action becomes more imperative as higher education demands more and more years from increasing numbers of intelligent young adults. The reward for the expense is more fine youngsters—and more real security for those whose financial contributions make the intelligent youngsters possible.[7]

If you have children or grandchildren in college who would marry and have youngsters during this most natural time, give them all the help you can. Do not feel that bright young men should refrain from marriage until they can support their wives. This view is a carry-over from an agricultural economy and it leads to a considerable amount of immorality on our campuses. Our materialistic society needs to recognize more fully that marriage is primarily a biologic matter, not an economic one. The young who would marry are instinctively right. It is the extended education which, though justified for the intelligent, is unnatural.

If you do not have at least five children which you will send to college, even though married, then select one or more intelligent couples and give them financial aid in having their family as they gain a full education.[8]

People should realize that discouraging their children from marriage while they gain an advanced education encourages the dying out of their family. College administrators should recognize the situation also and plan more dormitories for couples, more nurseries and other measures so that higher education, and par-

----

7 Younger people who are bright enough to have some capacity for success tend to emulate your success. On the other hand, those with poor capacity for success tend to want to seize the fruits of your success— and a great number of them are even willing to see you put out of the way during the seizure.

Facts, even the most disquieting, are not so dangerous when recognized and acted upon as when ignored.

8 The Foundation for Advancement of Man will help you locate such if you wish. See the final page in this book.

ticularly post-graduate education, will not tend to sterilize students. Education and having children, both highly important, should not be mutually conflicting.

Let states reduce the tuition in their colleges and universities for students having children. Such students are producing future taxpayers, while childless students are not. Students who are reproducing justify compensatory treatment.

If a people or a nation will successfully combine, in its intelligent young couples, abundant childbearing with the acquisition of higher learning it will become, or remain, forever great.

OPPORTUNITY B: CORPORATE RELATIONS

Corporations search the colleges for technically competent individuals and for men with executive potential. The hunting grows more intense and relatively less successful as the need grows greater.

"Today, for the first time in our history, we are running short of skilled technical brain-power. The broad-scale application of advanced technology has increased our consumption of technically trained persons to the point at which our available supply is no longer sufficient to meet our needs. The disparity between supply and demand has been emphasized by the officials of two hundred large companies which together employ well over half of all scientists and engineers in industrial research and development in the United States . . . the officials of more than half of these companies reported that shortages of technical manpower are either hindering present research activities or preventing desired expansions in their research and development programs."[1]

"The gap between the supply of creatively and intellectually able individuals in America and the demand . . . is constantly widening."[2]

"There are more jobs on the executive level to be filled, and fewer people to fill them."[3]

----

1 Brown, *The Next Hundred Years, op. cit.*, p. 31.
2 *Talent*, U.S. Dept. of Health, Education and Welfare Bulletin No. 34, 1963.
3 William F. Holmes, Executive Development Director, Lever Bros. Co., in *Wall Street Journal*, August 26, 1966, p. 1.

"... only 70 per cent of the occupational openings for engineers and scientists in industry is being filled. There are almost frantic recruiting competitions that take place on the campuses of technical schools every spring. . . . There are serious shortages in the medical and nursing professions, and there is every indication that these shortages will become even more severe."[4]

"... the realities today are that in many areas . . . the crucial bottleneck is brains, not money."[5]

It is increasingly clear that the limitations on corporate growth are less and less the available capital and more and more the available competence.[6] Corporations have come to generate, through retained earnings, more of the capital they need. *They can also come to generate more of the brains they need* by encouraging the birth of more bright children. Let farsighted corporations progress from the primitive hunting stage into the husbandman stage in securing the competent personnel they want. Let them set aside funds to enable brilliant scientists, stu-

------

4 John Weir, *The Next Ninety Years* (Pasadena: California Institute of Technology, 1967), p. 57.

5 Philip H. Abelson, *Science*, Vol. 140, No. 3574, June 28, 1963, p. 1365.

6 "The main shortage in our country is management talent." —Arjay Miller, *Fortune*, August, 1968, p. 31.

"There are openings for additional executives from nearly all fields of science and engineering—in policy-making or managerial jobs paying $15,000 or more a year.

"The corporations would like to hire one-seventh more organic chemists at executive levels than they have now, and one-eighth more industrial, electronic and mechanical engineers. They could use one-tenth more physicists, electrical engineers, chemical engineers and electromechanical engineers.

"In future years, corporations report, there will be a growing demand for all the types of executives now being sought, and some new fields will be added to the list." —*U.S. News and World Report*, Vol. XLV, No. 3, July 18, 1958, p. 75.

"There is an increasing need for peoples of the higher levels of capacity . . . ." —Victor Serebriakoff, *I.Q.—A Mensa Analysis* (Tiptree, Essex: Anchor Press, Ltd., 1965), p. 64.

"The supply of able people in each generation . . . appears to be very largely determined by hereditary factors. If this is true then the increasing supply which is needed could best be obtained . . . by a higher birth rate in those groups of the population who are above average in innate intelligence." —*Ibid.*, p. 70.

dents of engineering or of business management—whatever the corporation will need—to marry and to have children while completing their education.[7]

Corporations are excellently suited to accomplish this, for they are potentially as immortal as the governments which charter them. Enduring long, they can reap the benefits of long-range projects, such as an increase of good minds, to an even greater degree than individuals can.

Ours is a relatively free society, so that the offspring would not be commited to work for the corporation which helped to finance their existence. But the corporation which was instrumental in their coming into being would stand the best chance of securing their services when they were adult. As a matter of quicker return the outstanding fathers, with one or several children financed by the corporation, would themselves be inclined to join their benefactor when they finished college.[8]

———————

7 In the United States the laws permit 5% of corporate profits to be free of tax if properly expended for such purposes. —*U.S. Internal Revenue Code*, Sec. 170 (c) (2), 1954.

"The (President's) Committee . . . estimates that corporate contributions . . . for all charitable purposes was 1.27% of pre-tax income." —Laird Bell, *Engineering and Scientific Education Conference Proceedings* (Chicago: Oct., Nov., 1957), p. 53. Available from National Academy of Sciences, 2101 Constitution Avenue, N. W., Washington, D.C.

8 In the factories I manage we utilize a variant of this plan and limit it to our employees: Certain young men who are "comers," if they are married to an intelligent young woman and have at least one normal child, receive our offer to pay the prenatal and obstetrical care of each additional child they have. Also, depending on our estimate of the worth of the man, we offer to pay some or all of the cost of advanced education of his additional child or children. This last we handle through endowment insurance with the employee's children as beneficiaries.

I regard this as a wholly constructive plan. In my opinion, the larger a man's family the less is he inclined to move away. This is as effective in holding men as a deferred compensation plan. We have deferred profit-sharing compensation also; the two influences are mutually reinforcing. In addition, we have some good second-generation "dividends" coming along. We will have a preferred status as a potential employer with most of these.

If your own outstanding employees are not raising plenty of children who will be capable of succeeding them in time, you are missing one of the best ways for your corporation to grow through the years. I submit that "raising your own" or at least doing so in part, is at least as sensible as outbidding other companies for employees and thereby raising employment costs all around. We outbid others too, when neces-

OPPORTUNITY C: LIVING REMEMBRANCES

Many superior young couples can be encouraged to have more children by emphasizing that they are given only a certain time in the cycle of life in which to attend to the great matter of reproduction, and by sharing with them, if need be, the economic cost of additional children. The name of each additional child might well be that of the individual whose donation made the existence of the child economically feasible, or the name of a departed relative whom the donator wished to honor.

For example, Mrs. Jane Roe donates a substantial sum to a foundation,[1] in memory of John Roe. This sum will go far toward raising and educating a child today.

George Edwards, a young minister, and his wife have two superior children, but their budget is a lean one, and they do not feel able to increase their economic burden with more children. The foundation suggests to this young couple that if they will have an additional child and name it John Roe Edwards, the Roe bequest will defray much of the cost. Quite possibly within a year or so there will have come into the world a fine new youngster and John Roe will have been superbly memorialized for at least a lifetime.[2]

OPPORTUNITY D: LOOSENING THE CITY'S GRIP

Child Parks

The marked urbanization which is characteristic of the late stages of civilizations appears to be a major cause of thinning the

---

sary, but consider it necessary less often than if we did not have the above plan. For one thing, raising your own increases the supply of competent people. It represents basic solution of a problem rather than living with it as best we can while the problem worsens.

---

1 One such is the Foundation for Advancement of Man. Information about it is on the last page of this book.

2 If the Edwards child turns out to be a girl, the Foundation may use a bequest honoring a woman. Another superior couple is then approached on behalf of the Roe bequest.

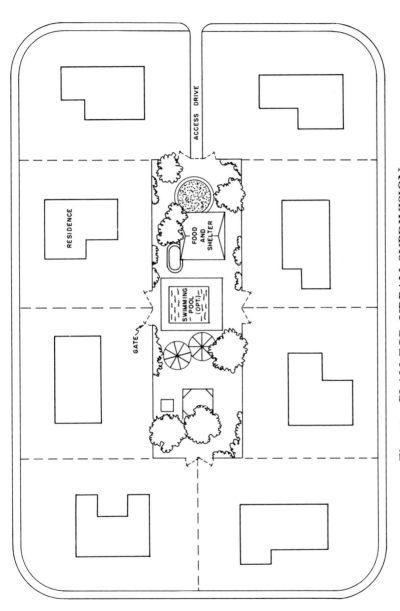

Fig. 10. PLAN FOR URBAN SUBDIVISION
Three-bedroom residences with fenced center for child play and care.

ranks of the intelligent strains which are attracted to cities.[1] In urban surroundings children are more of an encumbrance than elsewhere, hence their number tends to be severely limited among city couples who possess the intelligence to limit effectually the number of their children.

This gives rise to yet another approach to encouraging the increase of intelligent humans. There is an opportunity to create for intelligent young couples an environment which, unlike our present cities, is conducive to the having and rearing of children. In favorable areas in or near cities or near colleges, residence units could be constructed for occupancy by those who meet standards set up by the individual or group which financed the units. Such a development might well consist of assorted and attractive homes, each with at least three bedrooms.[2] These homes could surround a supervised child park and play area in which would be located a child care center. (See Fig. 10.)

In these homes the mothers, freed of child care whenever they wished, free to recuperate rapidly from child-bearing, surrounded by other women actively bearing children, would naturally be inclined to have more children than otherwise.

Occupancy within such subdivisions could be limited not only to those selected but to those who had a child at least every two years or so. Rentals could be nominal, thus easing the burden on the fathers. Being a relatively homogeneous and select group, such a subdivision would have marked social attractions, as well as exceptionally desirable physical accommodations.

Young fathers and mothers living in such an environment would be quite free to complete advanced educations or to pursue

---

1 ". . . there is a curious characteristic of the city which seems in all ages and all regions to have led to a lowered rate of reproduction. Mortality has perhaps often tended to be higher in cities than in the country, because of the increased hazards of disease; but fertility has also tended to be lower. It has been asserted that the great city has never reproduced itself, but has always depended on immigration from surrounding territory . . . ." —Marston Bates, *The Prevalence of People* (New York: Scribner's, 1955), p. 57.

2 But not large or pretentious homes. Luxury and pretentiousness appear to be hostile to the having of children. If a society were to take intelligent selection seriously, any young couple who lived luxuriously when, instead, they could be creating a full family, would be recognized as unnatural.

careers intensively. They could be with their children as much as circumstances and inclinations dictated and they could also be free of them whenever they needed to be. The adjacent child care center could attend to their children whenever the parents wished.[3]

Denominational colleges could have one or more such sub-divisions, for outstanding students and young faculty members, near their campuses. The money expended on such would increase the number of self-supporting and denomination-supporting members in the next generation.

Religious denominations which built such accommodations near large, non-denominational universities, for the use of their own married members who were students or young faculty there, would also be assured of a fuller crop of fine youngsters raised in their religion.

OPPORTUNITY E: LOOSENING THE CITY'S GRIP

Youth Farms

Farms are outstandingly the best environment for growing children. They are safer than the wilds of nature and more wholesome than cities. They are the most economical way to raise children and most children are happiest on them. They permit the outdoor physical activity which children need to develop fully their bodies and keep their emotional problems to a minimum.[1]

Youth farms, accessible to cities and schools, could be established on which youngsters could do much of their growing up. These need not be on the best land, for their most valuable crop would be the children. They should have cows, chickens, pasture, orchard, wood-lot, truck gardens and perhaps a hay field. They

————

3 Where means permitted, paid professional child care could be provided all or part time. Where means did not permit, the mothers could rotate in supervising the play and care area, so that all but one or two mothers could, if they wished, be free of child care at any one time. Fathers could participate on week-ends or holidays.

1 "The emotionally disturbed child is more likely to come from the city than the country, and many authorities cite urbanization as a factor contributing to this problem." —Barbara Culliton, "Children in Conflict," *Science News*, No. 90, August 20, 1966, p. 122.

should have a workshop and simple, fireproof accommodations for a moderate number of youngsters.

They would serve some of the same functions as private boarding schools, but be far less expensive because the youngsters would be in part self-sustaining and could utilize local public schooling. Children could visit their parents on week-ends, or alternate week-ends, and on vacations. Parents could visit their youngsters on the farms and even work with them there whenever they chose.

A major factor in the success or failure of a youth farm would be the couple or other supervisors chosen to manage it. This is true of any enterprise. There are many warm-hearted and industrious farm-owning couples who could accommodate, for a consideration, young guests from a nearby city. Or people could be selected to manage youth farms set up specifically for youngsters. If the youths admitted to a farm were chosen for the most important human qualities—health, intelligence and cooperativeness—the cost of supervision would be moderate. A youth farm should be better, yet less expensive, than raising the same youngsters in a city.

Plans D and E are not without their limitations and problems but neither are city living or suburban living combined with commuting. Youth farms are a way in which intelligent people, who now gravitate to cities and then cease to have many children, could combine city or suburban living with a greater and more normal number of bright children, and these children would be freed much of the time from the crowded and hazardous city.

OPPORTUNITY F: GERMINAL REPOSITORIES

In the United States approximately twenty thousand children are born each year as a result of artificial insemination.[1] The number is growing rapidly.[2] These are largely from inseminations of women who are unable to have children by their hus-

---

1 "There are over a quarter of a million A. I. children in the United States . . . many families boast of four or more inseminated children." —Wilfred Finegold, *Artificial Insemination* (Springfield, Ill.: Thomas, 1964), pp. 58, 63 and 111.
2 *Medical World News*, February 28, 1964, p. 74.

bands. They and their husbands prefer this method to child adoption for it gives the woman genuine motherhood, and thus complete consummation of her physiologic purpose in life. Furthermore, husbands are inclined to accept and love children borne by their own wives more than they are the adopted children of strangers.

It is estimated that there are from one million to five million[3] potentially fertile U.S. wives deprived of maternity by the infertility of their husbands. Probably the ratio is similar in other countries.[4]

In artificial impregnation the germinal father and mother never meet; a physician is the intermediary. If the donors of the germinal substance used were the most superlative men of our time, the women recipients would bear the very finest children of which they were capable. These would be healthier and more capable of scholarship or leadership than would the children of ordinary men. If the women recipients were also select, many children of high ability would be brought into the world who never would have been born otherwise.

But this method of relieving infertility, so potentially beneficial to man, is not yet taken to any significant degree. One reason is that few of our most superlative men are inclined to put themselves on call to a physician. Neither are they inclined to risk possible emotional and even personal involvement.

There is now a way to surmount these problems. It is possible to keep male germinal substance viable under deep refrigeration for years.[5] Donors may contribute the essential material at their convenience and with complete anonymity and detachment. Children may even be fathered by men long dead. Consider what it could mean to the scientific effort if now we could call into being another twenty or more children of Lord Ruth-

---

3 Milton Golin estimates more than one million. See "Paternity by Proxy," *The New Physician*, Vol. 11, No. 12, December, 1962, p. 426. M. Leopold Brodney is quoted as estimating five million. *Chicago Daily News*, November 14, 1963.

4 ". . . ten percent of couples of childbearing age are infertile . . . fifty percent of the childlessness is due to the husband." —Finegold, *Artificial Insemination, op. cit.*, p. ix.

5 J. K. Sherman, "Improved Methods of Preservation of Human Spermatozoa," *Fertility and Sterility*, Vol. 14, No. 1, Jan.-Feb., 1963.

erford or Louis Pasteur, all by select women candidates for artificial insemination. Consider the multitude of gains which we might have if many additional sons of Thomas Edison could be engendered. Imagine the greater defensive capability of any nation if it had, say, ten sons of General Arthur MacArthur.[6]

At least one institution to store the germinal substance of our most creative men is in operation.[7] The more there are of similar facilities the greater the number of individuals who will be able to take advantage of them. Repositories are not expensive and should be set up in every large metropolitan center. These would make it possible to bring into being many more offspring of our finest and most creative men. It could give us thousands —even millions—more of the very individuals who can contribute most to the strength and progress of the nation and of man. The production of genius might be multiplied several fold. It would permit the restoration in some measure of the ancient and beneficial situation in which the brightest men left the most children. This is one of the great opportunities for increasing the fund of human intelligence.

"This . . . has the potentiality of determining which . . . countries will have the mental power necessary for world leadership in science, the arts and political and economic development . . . ."[8]

OPPORTUNITY G: MILLIONS MAY PARTICIPATE

Let mature individuals repay the direct costs of childbearing for intelligent young couples who agree to have an additional

---

6 The father of General Douglas MacArthur.

7 Donors are at present limited to Nobelists who are relatively free of genetic defect. Other repositories could be limited to donors with exceptionally high intelligence and good general heredity.

Artificial insemination in humans was first suggested by Hermann J. Muller in 1908, in a paper presented before the Peithologian Society while he was a student at Columbia University. The first facility for preserving the genes of some of our finest men and supplying them for artificial insemination was set up in 1966 and named in honor of Dr. Muller. If information is desired regarding this, or regarding the establishment of similar repositories, write the Foundation for Advancement of Man, P. O. Box 2480-D, Pasadena, California 91105.

8 Weyl and Possony, *The Geography of Intellect, op. cit.*, p. 251.

child. If the particular situation calls for it, let the contributing individuals also take out sufficient endowment insurance, making the child the beneficiary, so as to provide for a college education whether or not the donators are still present when the child has reached college age.

Everyone with a little means and serious intent can utilize this straightforward approach. If you have bright children married to bright mates, do this for them. If you do not, and cannot find the right young couple or couples to be helped, the Foundation for Advancement of Man[1] can put you in touch with them.

_____

1 See end of book regarding F.A.M.

CHAPTER III

SUCCESS AT LIFE

Measures to encourage the birth of more children among the intelligent are important to the future of man and the several specific ways just suggested are only a fraction of the possibilities. Yet, surpassing even these in potential is the general attitude of intelligent people toward the having of children. This is a pivotal matter. The history of nations and the fate of man hinge upon it. If all people elected to have no children, mankind would be extinct within 100 years. If only the most intelligent elected to have children, the average intelligence of man would soar swiftly. If the attitude of the intelligent could be transformed so that spontaneously they would have more children, this would be a tremendously uplifting thing for all of mankind. There would be more wealth produced for all, including the masses. There would be greater control of disease and new sources of food, comfort, well-being and enjoyment. Accordingly, to intelligent and thoughtful individuals the following considerations are emphasized.

If you delay marriage or repress the number of children out of consideration for your parents, then remember that they did not give you love and care so that you would pay them back. They gave it so that you, in your turn, would give it to your children, even as their parents gave it to them. *Your debt is to the next generation, not to the previous one.* This is not a contractual debt, to be repaid to the lender; this is one which you owe to the potential children within you. Your parents gave you life, but you cannot give it back to them. You can only give it to your children in turn.

Competent young people today are in such demand and have such opportunities open to them that many more should be brought into the world to take advantage of the greatest fullness of life yet experienced by man. For the first time in history, a new baby can expect to live seventy or more years and can enjoy the greatest freedom from disease and the highest living standard ever known to man. He can not only expect to grow up but, in addition, he can see more of the world, labor less and experience more exciting extensions of his own powers than ever before. If you can have superior children and there is a question as to whether or not to have more—decide affirmatively. Since we may expect to live longer than the generations who came before us, and since children are the greatest interest, comfort and protection one may have in later life, they are the finest investment one can make while young. They will enrich your life and be among the very greatest satisfactions of your lifetime. You can, while young, increase the number of those who will still love you in your old age. Nothing else on which you expend effort and resources can bring such deep and abiding joy as a full family. If you are intelligent enough and thoughtful enough to have read thus far, there should be many more of your kind. The world needs more of what you can add to it.

There is needed in the free nations a greater proportion of men of good will and clear-minded leadership. There is needed a greater proportion of women who perceive the need for and who teach—largely by example—integrity, goodness, love of fellow man—principles about which there is no argument, only a great lack.

If you can give to your children "the most lasting, the most persistently satisfying, the most all around useful of natural endowments—a really good mind"[1]—then give this gift abundantly. It may be given again and again while your store of it remains undiminished. This is "the cruse which never runs dry." It is more precious by far than Aladdin's lamp, if not so instantaneous. In your own circle many may seem to possess good minds but this is because you tend to associate with others like yourself. Actually, relatively few possess this gift. If you are one who does, multiply your gift generously. Increase the number of good minds in your own family and, so doing, in your church, your

---

1 H. G. Rickover, "The Talented Mind," *Science News Letter,* March 16, 1957, p. 170.

city and your nation. " . . . the quality of any society is directly related to the quality of the individuals who make it up."[2]

Today the impulse to exhibit the outward evidences of material success is so excessive that it drives some very able men to burn themselves out and women to limit their families so as to help their men strive and make a showing. Now that economic success is heavily penalized by income taxes, such a fetish is an unrealistic carry-over from a pioneering economy.[3] There are many forms of success other than acquisition of money, influence or fame. There are successful innovators, constructors, discoverers, inventors, researchers and many others—but the greatest of all successes, and the one most enduring and satisfying, is success at bringing into being many other humans as good or better than ourselves. All persons may accurately gauge their success, and the success of others, at recreating life itself. The following considerations are basic.[4]

The number of children a couple may have varies from none to a possible 69.[5] Neither of these extremes represents the ideal. For an intelligent couple to have no children results in the genetic death of the precious genes for intelligence of which the couple are the living custodians. For an intelligent couple to have as many children as possible would indeed maximize their genetic contribution to the next generation, but this is not necessary and in many cases would impair the environmental advantages (especially the education) which the parents might give to a mod-

---

2 Miller Upton (President of Beloit College), *U.S. News and World Report,* July 10, 1967, p. 63.

3 Let our society pay less attention to conspicuous possessions as a way of demonstrating economic success. Instead, let the plurality of intelligent children be recognized as the badge of complete success. Let there be less emphasis on pecuniary success and more on success at life itself.

4 In saying these things I may hurt many dear friends. Yet if these few basic facts were to be more widely realized, this alone could change the future history of man. And I mean no hurt. What I really want is more like them.

5 "The remarkable mother who bore the five famous Rothschild geniuses had 20 children." —Frederic Morton, *The Rothschilds* (Greenwich: Fawcett Publications, 1961).

". . . an Austrian woman, Mrs. Bernard Scheinberg . . . bore sixty-nine children. . . ." —Bates, *The Prevalence of People, op. cit.,* p. 83.

erate number of children. Moderation usually proves best. The following considerations are important in arriving at what constitutes responsible moderation.

No Children

Obviously, one who fails to reproduce is not a complete success at life, although he may be successful in other ways, and even contribute importantly to others' success at life. Nevertheless, for himself or herself the life-score stands at zero. Whosoever fails to hand on the torch of life—handed down to him in unbroken succession since the beginning of life on earth—destroys his potential descendants for incalculable generations. He becomes an evolutionary dead end. This is nature's unforgivable sin. All who commit it and all of their particular kind are eradicated forever. In innately stupid or similarly unfortunate individuals this is not a great tragedy, for biologic regression is already evident in them. But among superior humans, of all the sins of omission, none is so grave as the willful termination of that long and wonderful continuum of lives which led to their own existence; as the ending of that precious chain of ever-renewable intelligence of which they are the terminal link. They, the repository of some of the most precious of human genes, are a faithless repository. That portion of the human gene-pool of which they were the sole custodians has been lost irretrievably. In that loss mankind has died a little.[6]

"Every intelligent person who appreciates and enjoys the cultured life of a hard-won civilization should, moreover, reflect

––––––––
6 Every loss of genes for high intelligence is a loss of the uniqueness of man, an erosion of the essence of humanity. The childlessness of an Isaac Newton or a George Washington, the extinction of the Lincoln family, the spinsterhood of the brightest girl in the class, are losses to all mankind. They mean that man has lost a little of that essential quality which separates him from the beasts.

". . . when we study the reproductivity of those who stand out above the herd. . . , we find that it is very low compared with the reproductivity of those who segregate at the other end of the scale. The trend is obvious. For awhile, . . . there will be good genes pervading the germ-plasm of the general population awaiting combination into the superior individual. But one cannot watch the extinction of this germ-plasm whenever it does result in a high-class human product without realizing that a tragedy is being enacted." —Edward M. East, *Heredity and Human Affairs* (New York: Scribner's, 1927), p. 264.

that by his failure to breed, in the face of a rising tide of defective intelligences, he brings a relapse into brutality one step nearer."[7]

"Charles Darwin's grandfather was Erasmus Darwin, physician, poet and philosopher, and independent expounder of the doctrine of organic evolution. Darwin's father was a distinguished physician, described by his son as 'the wisest man I ever knew.' Darwin's maternal grandfather was Josiah Wedgewood, the famous founder of the pottery works. Amongst his first cousins is Sir Francis Galton. He has four living sons, each a man of great distinction, including Mr. Francis Darwin and Sir George Darwin, both of them original thinkers, honored by the presidency of the British Association. No one will put such a case as this down to pure chance or to the influence of environment alone. This is evidently, like many others, a greatly distinguished stock. The worth of such families to a nation is wholly beyond anyone's powers of estimation. What if Erasmus Darwin had never married!"[8]

Bright, healthy and fertile individuals who deliberately have no children are not to be regarded as smart, but as forfeiting their birthright. They deny life to the fine, intelligent children within them. In future, less-confused societies they may well be considered criminally delinquent.

If failure to realize these facts has delayed your biologic success, then know that the most adorable things in the world are babies, and the most fascinating babies are your own. Exert every effort to have these renewing extensions of your own life before the magic of being able to create life passes from you. But if circumstances have permanently robbed you of biologic success, encourage and help other worthy ones to achieve it.[9]

One Child[10]

For two superior adults to have but one child is so inadequate a performance as to need little comment. If it is impossible for you to have more than one child, then the hearts of those

---

7 Cattell, *The Fight for Our National Intelligence, op. cit.*, p. 139.
8 Saleeby, *Parenthood* . . . , *op. cit.*, p. 336.
9 The Foundation for Advancement of Man can assist in this if you wish. Information about it is on the last page of this book.
10 One child is a distraction; five children are a career.

more fortunate go out to you and to your only child. However, if you deliberately have but one child when you might have more, you deserve the scorn of all thoughtful people.[11] You are delinquent in your most important office, a partial defaulter of the trust of life.

Two Children

Two children are a good beginning for a young couple. Yet for two adults to produce but two children does not fully repay the debt to nature or replace themselves, since not all children live to maturity and reproduce. Three are the fewest a couple can have without diminishing their kind.

Until a couple fully replace themselves, they are biologically deficient. The purpose of their union is partially served but has not been fulfilled. They may satisfy their sexual and parental urges but they have not paid their biologic debt.[12] Let them not stop at two but give the gift of life at least as abundantly as it was given to them.[13]

If you can have children with intelligence above the average yet you contemplate stopping with only one or two, consider what you are doing. You may attain a higher socio-economic status than otherwise, but you will reach it by stifling the lives of the several bright children whom you could have brought into the world.

Are you willing to do that?

Some are actually willing, perhaps because they do not realize that they attained their status by cheating on society. And

---

11 "The thoughtful young citizen may well avoid the 'only' daughter as he would the plague!" —Cattell, *The Fight for Our National Intelligence, op. cit.,* p. 156.

"The extremely small family is no longer as 'fashionable' as it was. . . . Much has been written in recent years of the disadvantages of the one-child family both for child and parent. . . ." —*Great Britain Royal Commission on Population Report* (London: His Majesty's Stationery Officer, 1949), p. 56.

12 "Childless women and mothers of extremely small families have shorter expectation of life than mothers of moderate sized families." —Powys, in Samuel I. Holmes, *Trend of the Race* (London: Constable, 1921), p. 189.

13 A society which awards full social approval to childlessness or inadequate maternity among its brighter women has need to reappraise its mores, for it harbors the seeds of its own decline and engulfment.

184 OPPORTUNITIES OF THE INTELLIGENT

the few children they do have are coming to be recognized as less desirable mates than their social equals because children of poor child-bearers are themselves inclined to be feeble reproducers.[14] Do not let biologic failure be the example you set for your children.

If circumstances have limited you to two children and it is impossible to have more, then encourage and assist your children, in their time, to raise your family's score.

Three Children

Three children are the fewest a couple may have and still not contribute to a decline in human numbers. Three children qualify the parents as biological successes. However, three represent numerical success only. It would suffice if the population were static. But the population is increasing rapidly. Relatively they are not even equal to the average.

We easily see that a couple which has but one child diminishes their kind absolutely. We should also realize that a couple which has less than four diminishes their kind relatively when the general population is doubling every thirty-three years.

Four Children

Four is the least number of children a couple may have and still contribute their share to the growing population.[15] Four children represent average performance—nothing to be ashamed of, yet nothing exceptional. Most families in the world do at least as well. For those who would have better than average success at life, four is a good start. Exceptional success is still ahead.

Five Children or More

If a couple would do better than the average today, five children are the fewest which will let them qualify. Only then may they look all others in the eye with the calm, deep realization that, when judged by the most fundamental of standards, they are more successful human beings than most.

---

14 "Families with only one or two children die out rapidly, whereas families with five or more children double in each generation." —Osborn, *Preface to Eugenics, op. cit.*, p. 197.

15 "One for Mother, one for Father, one for accidents and one for increase." —Sir Winston and Lady Churchill, quoted by Elizabeth Nel in *Mr. Churchill's Secretary* (New York, Coward-McCann, 1958).

This will be only a part of their satisfaction. It is one of the miracles of children that each added child is on the whole markedly less trouble than the last; yet each is an experience almost as full of new wonder and joy as the first. The creation of fine, new life retains its wonderment, while the ability to manage children, and the ability to love and enjoy them, increase as the family increases.[16] The more children there are, the more the older ones help the young, and the fuller is the whole life of each child, as well as the lives of the parents.

A family of five children represents for an intelligent couple responsible moderation in parentage.[17] It is certainly not an upper limit. The greater the endowment of intelligence which a couple can bestow upon their children, the more children they should have. The more they multiply their own intelligence the more there is in their family and in the world. This is the miracle which compares even with that of the loaves and fishes, yet every intelligent couple may perform it.[18] If you can bring more in-

---

16 ". . . it is much easier to raise several bright children than one or even two, as well as much healthier for the individual child. . . ." —Dora Crouch, mother of six, all with I.Q.'s above 140.

17 The figure of five as the minimum number of children per couple for outstanding biological success is also arrived at if we take a quite different approach. It was said by Galton and is true in a special sense that the parents' contribution to the genetic constitution of their children is about one-half (the other 50% being contributed by earlier ancestors). —Sir Francis Galton, *Natural Inheritance* (New York: Macmillan, 1889), p. 136.

"The average genetic correlation between parent and child . . . is 0.50." —L. Erlenmeyer-Kimling and Lissy F. Jarvik, "Genetics and Intelligence," *Science*, Vol. 142, No. 3598, December 13, 1963.

It thus takes four children for a couple to leave the equivalent of their own hereditary characteristics in the next generation if all four children live, and if the rule of one-half has worked out in their particular case. (This fraction, being statistically arrived at, will not suffice for 49% of the population.) One additional child is the least that can be produced to allow for these two contingencies. Accordingly we arrive again at a figure of five as the fewest children an intelligent couple may have without probably contributing to a diminution in the next generation of their own special genetic qualities.

18 ". . . psychologists have established beyond doubt that intellectual capacity is inherited. Like all other inheritable traits, it may come from distant ancestors rather than from the immediate parents, so that college professors sometimes have dull sons and stupid parents sometimes have

telligence into the world than most can, do so abundantly.[19] *The more intelligent you are the more children you should have.*[20] Full realization of the transforming potential of this principle can lift humanity above its present problems and limitations. It can restore to man the essential condition which, in the past, resulted in his ascendance and which, in the future, can give him even greater progress.

"As far ahead as I can see, there is no barrier to our evolution but man is at a crisis and must act quickly if he is to survive."[21]

---

brilliant children, or the same family may have one bright, one average and one dull child. Nonetheless it works out by and large that brighter parents tend to have brighter children. The children of professional men have been shown to have the highest I.Q.'s (averaging around 115) and those of day laborers the lowest (around 96). A child born into a poor home often shows an improvement in I.Q. if adopted by a more intelligent and stimulating family. . . . But the amount of improvement is always limited by the mental capacity that was there at birth." —"The Meaning of Psychology," *Life Magazine*, Vol. 42, January 14, 1957, p. 118.

The variation in intelligence of their offspring will be as great for the bright as for the dull, but with the bright it centers about a higher level. Thus, the more intelligent the parents the more intelligent the children, on the average.

"No breeder doubts how strong is the tendency to inheritance; that like produces like is his fundamental belief: Doubts have been thrown on this principle only by theoretical writers." —Darwin, *On the Origin of Species, op. cit.*, p. 19.

"Superior children come from superior lines, or at least from good lines. Geniuses are never the offspring of mentally deficient parents. Likewise feeble-minded children come largely from stocks of inferior intelligence. They are not produced at random from the general population.

"A relatively few family lines furnish a relatively great proportion of the eminent people of our country, or of any other country." —Snyder, *The Principles of Heredity, op. cit.*, p. 331.

19 No power that we have to shape our own nature or that of our fellows can hope to equal the immense power we hold over the nature of our own descendants. It is ours to say whether they shall come alive or be denied life: whether they shall be many or few and, depending in part on our choice of mate, whether they shall be taller or shorter, stronger or weaker, brighter or duller than we.

20 If the mean I.Q. of you and your mate is over 100, have at least five children. (Intelligence Quotient is not a perfect gauge of intelligence but is as good as we have.) Likewise, if the mean I.Q. of you and your mate is over 110, have at least six children. If it is over 120, have seven children; over 130, have eight; over 140, have nine; and so on.

21 Hermann J. Muller in address to 3rd International Congress of Human Genetics, University of Chicago, 1966.

Those who are poor reproducers usually rationalize their position with thinking based essentially on fear or selfishness. But those who have children adequately need no resort to rationalizations. They sense their affinity with nature, their fulfillment of their life's purpose, which the poor reproducers will never know. The sleeker car or more expensive neighborhood which the biologically deficient can afford—since they spend most of their substance on themselves—are only fleetingly envied by those who have the deeper and more fundamental satisfactions of a goodly family.

"History is full of regrets expressed in old age by men who had sought fame or fortune and felt that their years had been misspent. We can not recall ever having heard of a man who, after bringing up successfully a family of superior children, expressed regret over a waste of opportunities."[22]

In our society, which is thinning at the top though it need not,[23] some of the attitudes toward ample families have become so irrational that it is well to point out that a normal young woman who marries five years after becoming fertile, and who does not restrict her childbearing, can expect to be the mother of thirteen live children.[24] Fifteen is the normal complement if she marries early. For a healthy couple to fall far short of the normal number betrays some unnatural restraint upon the woman's reproductive faculties. Let the attitudes toward size of families be based, not on shifting opinions as to what is a stylish number but on the facts of nature. The five children per couple which today constitute success at life is the minimum for such success. A couple with a really large family leads a much fuller life than one with only five children.

There is no deeper or more abiding joy than that which comes from having fine, intelligent children. Only those to whom the experience has been denied might fail to recognize this. It

----

22 Paul Popenoe and Roswell Hill Johnson, *Applied Eugenics* (New York: Macmillan, 1933), p. 265.

23 "Societies may be differently successful . . . and when both sexes set their heart against reproduction, then such societies die out—even without benefit of contraceptives."   —Margaret Mead, *Male and Female* (New York: New American Library, 1949), p. 181.

24 Alan F. Guttmacher, "Factors Affecting Normal Expectancy of Conception," *Journal of the A.M.A.*, 855-860, Vol. 161, June 30, 1956.

is a joy which begins at the first birth, grows with the unfolding of the children's abilities and enriches the later years beyond all else. Would it not be good to raise this joy to its quintessence and spread it more widely throughout the lands, and in so doing benefit not only the parents, but the community and perhaps the world? If you are of superior endowment, yet you do not have at least five children, get you to work![25] If direct success at life is unattainable, then help each of your bright children to have as many more than five as you are short of that number.[26]

This will mean encouraging and enabling them to have children while they are young and probably while they are in college. And if that does not fully succeed, then for the sake of the intelligence of man and your own security, see to it that some other intelligent young couple comes to have as many more children above five as you are short of that number.[27]

This is a program of action for the increase of intelligence. This is a plan to enrich the biologic endowment of humankind.

----

25 Childbearing is far easier and safer than it was even fifty years ago. If the mothers of our nation when it was young could average eight children, five or more is indeed a modest goal. The safety of mothers at childbirth has increased drastically in the U.S. from a mortality rate of 64 per 1000 live births in 1932 to a low of 4.5 in the four years ended in 1958." —*Science News Letter*, May 14, 1960, p. 318.

26 A couple with one intelligent child should assist that child, when married, to have 5 plus the 4 the couple didn't have (9 in all).

A couple with 2 children should assist each child to have the basic 5 plus 1 or 2, to total 13 grandchildren.

A couple with 3 children should assist each child to have 5 and at least 2 of their children to have $6 = 17$ grandchildren.

A couple with 4 children should assist each child to have 5, and one of their children to have 6. $(3 \times 5) + 6 = 21$.

A couple with 5 children have done exceptionally well but can try for the physiologically normal number.

Imagine what a mighty increase there would be in the number of good minds at work on the problems of the world if this program were substantially realized among the highly intelligent!

27 This is how, for the intelligent who are past the age of parenthood, the quota for a finer future is arrived at, based on the principle of exceptional success at life: A childless person, single and past the age of parenthood, should help an intelligent couple to have at least 2½ more children than they would otherwise have had. This is arrived at mathematically. Obviously three is the actual minimum. A childless couple should help a young and intelligent couple to have 5 more children than the younger couple would have had without help.

Those with more money than their needs require can find no more constructive employment for their surplus. Those with time to devote can have no more fascinating avocation than encouraging this purpose. The whole development of man says clearly that his supreme purpose in life is to evolve upward in intelligence. This was the chief means by which he rose above the animals and escaped from a brute existence. It is the chief means by which he can rise above his present problems. Participating in constructive procreativity is the most meaningful thing we can do.

The creative response can produce, by thousands and millions, young, intelligent, trustworthy allies to share in the oncoming desperate stand and bring about, not merely a thwarting of the efforts of incited masses to extinguish your kind, but a greater goodness of living, a higher level of existence, a nobler purpose than man has yet known. It is so simple that it takes only a change in attitude to bring it about. Attitudes are subtle but powerful matters. They make the difference between love and hate, between peace and war, between a goodly family and barrenness. A false attitude can carry us to destruction, or if changed, unleash tremendous forces from within us. If the change comes slowly, as such changes often do, it may be too late. If it comes rapidly—and the urgency of our situation may bring it about rapidly—then the unleashing can lead us soon to a greater inherent strength and a higher plane than man has ever known.

How can this be achieved? We can make fuller use of our potentialities for bringing fine-minded people into the world to occupy it with us. We can in this way take the first and most essential step toward resuming our evolution into ever more intelligent beings. Let it be repeated, for it is the most potentially powerful fact in the affairs of man since he left the caves: *We can resume our evolution into more intelligent beings.* Man's inborn equipment for life can be progressively improved.

We have been given sufficient knowledge to do this. It takes only the catalyst of willingness in order to accomplish it.

Will the intelligent allow themselves to become increasingly outnumbered and exploited? Will most of them be liquidated everywhere, even as they have been throughout the great areas of the world where mass revolutions have prevailed? Or will the resurgence of the intelligent be such as to bring innate progress to man and achieve for all a level of life beyond any even dreamed of today? What a magnificent alternative to the present ominous situation, even if the possibilities be only partly achieved! Let us so strive that man will resume his evolution into an ever more intelligent creature; that today's awful threats be overcome; that our tomorrows become more illumined by reason and thought; that our world be increasingly filled with better, brighter, more decent peoples. At this living moment the future begins. What it shall be depends in significant measure upon you and what you do as you lay down this book.

BIBLIOGRAPHY

Abelson, Philip H., *Science*. Vol. 140, No. 3574, June 28, 1963.
Alison, Sir Archibald, *History of Europe*, Vol. I. Edinburgh and London, Blackwood, 1849.
Alvarez, Walter C., *Modern Medicine*. March 11, 1968.
Asia. June, 1924.

Bajema, Carl Jay, "Estimation of the Direction and Intensity of Natural Selection in Relation to Human Intelligence," *Eugenics Quarterly*, Vol. 10, No. 4, 1963; Vol. 13, 1966; and Vol. 15, No. 3, 1968.
Balk, Alfred and Harley, Alex, "Black Merchants of Hate," *Saturday Evening Post*. Vol. 236, January 26, 1963.
Bates, Marston, *The Prevalence of People*. New York, Scribner's, 1955.
Beadle, George W., "The Uniqueness of Man," *Science*. January 4, 1957.
————, *Physical and Chemical Basis of Inheritance*. Eugene, Oregon State System of Higher Education, 1957.
Beard, Charles, *Whither Mankind*. London, Longman's, 1934.
Bell, Laird, *Engineering and Scientific Education Conference Proceedings*. Washington, D.C., National Academy of Sciences.
Bendar, Clemans E., *Journal of the American Medical Association*. Vol. 201, July 17, 1967.
Berrill, Norman J., *Man's Emerging Mind*. New York, Dodd Mead, 1955.
Bloomfield, Paul, *Uncommon People*. London, Hamilton, 1955.
Bogue, Donald I., *The Population of the U.S.* Glencoe, Ill., Free Press, 1959.
Bonner, Jas., *The Bulletin*, January 21, 1965, Pasadena, Calif. Institute of Technology.
Boule, Marcellin and Vallois, Henri, *Fossil Men*. New York, Dryden, 1957.
Braidwood, Robert J., "Near Eastern History," *Science*. Vol. 127, June 20, 1958.
————, *Prehistoric Man*. Chicago, Natural History Museum, 1959.
Brinton, Crane, *The Anatomy of Revolution*. New York, Knopf, 1952.
Brown, Harrison, *The Challenge of Man's Future*. New York, Viking, 1954.
Bruckberger, R. L., "A 2nd U.S. Revolution," *Life*. July 13, 1959.
Burke, Edmund, *Reflections on the Revolution in France*. New York, Liberal Arts Press, 1955.
————, *Science*. Vol. 160, May 10, 1968.
Burt, Sir Cyril Lodovic, *Intelligence and Fertility*. London, Hamilton 1946.
————, Introduction to *The Future of Man*. N. Quincy, Christopher, 1970.

Carrell, Alexis, *Man the Unknown*. New York, Harper, 1939.
Carr-Saunders, Alexander M., *World Population: Past Growth and Present Trends*. London, Oxford, 1936.

Cattell, Raymond B., *The Fight for Our National Intelligence*. London, King, 1937.

Childe, V. Gordon, *What Happened in History*. Baltimore, Penguin, 1964.

Churchill, Winston, *Commons*. March 28, 1950.

Clark, F. and Synge, R. L. M., eds., *The Origin of Life on the Earth*. New York, Pergamon Press, 1959.

Clark, Fred G. and Rimanoczy, Richard S., *Why Communists Hate—*. New York, American Economic Foundation.

Clark, Grahame, *From Savagery to Civilization*. London, Cobbett, 1946.

Clark, Sir W. LeGros, *History of the Primates*, 3rd ed. London, British Museum, 1953.

Cleland, Herdman F., *Our Prehistoric Ancestors*. Garden City, Doubleday, 1928.

Colin, Edward C., *Elements of Genetics*, 2nd ed. New York, McGraw, 1946.

Collie, George L., "The Aurignacians and Their Culture," *Beloit College Bulletin*. Vol. 26, 1928.

Conquest, Robert, *The Great Terror*. London, Macmillan, 1968.

Cook, Robert C., *Eugenics Quarterly*. September, 1965.

—————, *Human Fertility*. New York, Sloane, 1951.

Coon, Carleton S., "Living Stone Age Tribe," *Science News Letter*. Vol. 70. September 15, 1956.

—————, *Story of Man*. New York, Knopf, 1954.

Crow, James F., "Possible Consequences of an Increased Mutation Rate," *Eugenics Quarterly*. Vol. 3-4, June, 1957.

—————, "Mechanisms and Trends in Human Evolution," *Daedalus*. Summer, 1961.

—————, *Scientific American*. Vol. 201, No. 3, September, 1959.

Culliton, Barbara, "Children in Conflict," *Science News*. No. 90, August 20, 1966.

Curle, J. H., *Our Testing Time*. New York, Doran, 1926.

d'Abro, A., *The Evolution of Scientific Thought*. New York, Dover, 1950.

Dallin, David J., *The Real Soviet Russia*. New Haven, Yale, 1947.

Dampier, Sir William, *A History of Science*. London, Cambridge, 1966.

Darwin, Charles Robert, *Variation of Animals and Plants Under Domestication*. New York, Orange Judd, 1868.

—————, *The Descent of Man and Selection in Relation to Sex*. New York, Appleton, 1871.

—————, *The Descent of Man*. New York, Modern Library.

Darwin, Sir Charles, *Problems of World Population*. Cambridge, Cambridge University Press, 1958.

—————, *The Next Million Years*. Garden City, Doubleday, 1952.

Deevy, E. S., *Scientific American*. 1960.

Djilas, Milovan, *Conversations With Stalin*. New York, Harcourt, 1962.

Dobzhansky, Theodosius, *The Biological Basis of Human Freedom*. New York, Columbia University, 1960.

————, *The Biology of Ultimate Concern.* New York, New American Library, 1967.

————, *Evolution, Genetics and Man.* New York, Wiley, 1955.

————, "Genetic Loads in Natural Populations," *Science.* Vol. 126, August 2, 1957.

————, *Mankind Evolving.* New Haven, Yale, 1962.

Dodson, Austin I., *Synopsis of Genitourinary Diseases.* St. Louis, Mosley, 1952.

DuBridge, Lee A., "The Challenge of Sputnik," *Engineering and Science.* Vol. XXI, February, 1958.

————, *Education in the Age of Science.* New Haven, Yale, 1958.

Dublin, Lewis, *Fact Book on Man.* New York, Macmillan, 1963.

Dulles, Allen, *The Craft of Intelligence.* New York, New American Library, 1965.

Dunn, L. O. and Dobzhansky, Theodosius, *Heredity, Race and Society.* New York, New American Library, 1952.

Durant, Will and Durant, Ariel, *The Lessons of History.* New York, Simon & Schuster, 1968.

East, Edward M., *Heredity and Human Affairs.* New York, Scribner's, 1927.

Einstein, Albert, *Ideas and Opinions.* New York, Crown, 1954.

————, *The World as I See It.* New York, Philosophical Library, 1949.

Elliott, H. Chandler, *The Shape of Intelligence.* New York, Scribner's, 1969.

Ellis, Havelock, *More Essays of Love and Virtue.* Garden City, Doubleday, 1931.

————, *The Task of Social Hygiene.* New York, Houghton Mifflin, 1912.

Encyclopedia Britannica, 14th ed., "Brain, I"; "Industrial Revolution"; "Russia."

Erlenmeyer-Kimling, L. and Jarvik, Lissy F., "Genetics and Intelligence," *Science.* Vol. 142, No. 3598, December 13, 1963; *Eugenics Quarterly,* Vol. 13, Sept., 1966.

Finegold, Wilfred, *Artificial Insemination.* Springfield, Ill., Thomas, 1964.

Fisher, Sir Ronald, *The Genetical Theory of Natural Selection.* New York, Dover, 1958.

Flannery, Kent V., "The Ecology of Early Food Production in Mesopotamia," *Science.* Vol. 147, March 12, 1965.

Folk, Hugh, *The Shortage of Scientists and Engineers.* U.S. Dept. of Commerce, 1968.

Fortune. Vol. 70, August, 1964.

Galton, Sir Francis, *Hereditary Genius.* New York, Horizon, 1952.

————, *Memories of My Life.* New York, Dutton, 1908.

————, *Natural Inheritance.* New York, Macmillan, 1889.

Gantt, W. Horsley, "Pavlov, Champion of the Truth," *Modern Medicine.* November 12, 1962.

Gaxotte, Pierre, *The French Revolution.* New York, Scribner's, 1932.

Girard, R. W., "Intelligence, Information and Education," *Science.* Vol. 148, May 7, 1965.

Golder, F. A., *The Lessons of the Great War and the Russian Revolution.* Palo Alto, Cal., Stanford University, Hoover Library, 1924.

Golin, Milton, "Paternity by Proxy," *The New Physician.* Vol. 11, No. 12, December, 1962.

Good, William, *World Review and Family Patterns.* New York, Free Press, 1963.

Gosselin, Louis Leon Theodore, *Paris in the Revolution.* New York, Brentano's, 1925.

Grabill, Wilson H., *Fertility of American Women.* New York, Wiley, 1958.

Grass Roots Forum. Vol. 3-1, May 30, 1969.

Greenwall, Constantin de, *The Churches and the Soviet Union.* New York, Macmillan, 1962.

Grey, Ian, *The First Fifty Years,* New York, Coward-McCann, 1967.

Griffing, John B., "A Comparison of the Effects of Certain Socio-Economic factors, *Journal of Heredity.* Vol. XXXI, 1940.

Guttmacher, Alan, "Factors Affecting Normal Expectancy of Conception," *Journal of the A.M.A.* Vol. 161, June 30, 1956.

Haldane, J. B. S., "Human Evolution," *Genetics, Paleontology and Evolution.* Princeton, Princeton University Press, 1949.

————, "Science and the Future," *Daedalus.* 1924.

Harrison, George R., *What Man May Be.* New York, Morrow, 1956.

Harvard Business Review. July, 1957.

Harwood, E. C., *20th Century Common Sense and the American Crisis of the 1960's.* American Institute for Economic Research, 1960.

Hauser, Philip M., "Demographic Dimensions . . .," *Science.* Vol. 131, June 3, 1960.

Hazlitt, William, *Life of Napoleon Bonaparte,* Vol. I, New York, Wiley & Putnam, 1847.

Herald, J. C., *The Mind of Napoleon.* New York, Columbia, 1955.

Higgins, Edward Leroy, *French Revolution.* Boston, Houghton Mifflin, 1939.

Higgins, J. V., E. W. Reed and S. C. Reed, "Intelligence and Family Size." *Eugenics Quarterly,* Vol. 9, 1967.

————, *The Eugenic Predicament.* New York, Macmillan, 1933.

Holmes, Samuel J., *Studies in Evolution.* New York, Harcourt, 1923.

————, *Trend of the Race.* London, Constable, 1921.

Holmes, William F., *Wall Street Journal.* August, 26, 1966.

Hoover, J. Edgar, *A Study of Communism.* New York, Holt, 1962.

————, *American Mercury,* Vol. 86, January, 1958.

————, *Masters of Deceit.* New York, Holt, 1958.

————, *On Communism.* New York, Random House, 1969.

Howell, F. Clark, *Early Man.* New York, Time-Life, 1968.

Howells, William, *Mankind in the Making.* Garden City, Doubleday, 1959.

————, *Mankind So Far.* Garden City, Doubleday, 1944.

Huntington, C. C., *The Geographic Basis of Society*. New York, Prentice-Hall, 1933.

Huntington, Ellsworth, *Mainspring of Civilization*. New York, Wiley, 1944.

Huntington, Ellsworth and Whitney, Leon F., *The Builders of America*. New York, Morrow, 1927.

Huxley, Sir Julian, "Eugenics in Evolutionary Perspective," *The Eugenics Review*. October, 1962.

————, *Evolution in Action*. London, Gollanz, 1953.

————, *Evolution in Action*. New York, Harper, 1953.

————, *Man in the Modern World*. London, Chatto, 1947.

Intelligence Digest. Vol. 24, February, 1962.

————. Vol. 26, April, 1964.

————. No. 345, August, 1967.

————. No. 350, January, 1968.

Jepson, Simpson and Mayr, *Genetics, Paleontology and Evolution*. Princeton, Princeton University Press, 1949.

Johnson, Ellis A., *U.S. News*. January 31, 1958.

Jordan, David Starr, *The Human Harvest*. Boston, Beacon, 1907.

Kallman, Francis J., *Medical Tribune*, April 3, 1964.

Keith, Sir Arthur, *Ancient Types of Man*. New York, Harpers, 1918.

————, *Concerning Man's Origin*. London & New York, Putnam's Sons, 1928.

————, *Evolution and Ethics*. New York, Putnam, 1947.

Ketchum, Richard M., ed., *What Is Communism?* New York, Dutton, 1963.

Khrushchev, N., *Combined Reports on Communist Subversion*. U.S. Government Printing Office, 1965.

————, *Kommunist*, No. 12, 1957.

Korol, Alexander G., *Soviet Education for Science and Technology*. Cambridge, Mass., 1954.

Kramer, Samuel Noah, *History Begins at Sumer*. Garden City, Doubleday, 1959.

Kroeber, Alfred L., *Anthropology*. New York, Harcourt, 1958.

Kroeber, A. L. and Waterman, T. T., *Source Book in Anthropology*. New York, Harcourt, 1931.

La Barre, Weston, *The Human Animal*. Chicago, University of Chicago Press, 1954.

Langdon-Davies, John, *The Seeds of Life*. New York, New American Library, 1955.

Lansner, Kermit, *Second-Rate Brains*. Garden City, Doubleday, 1958.

Lassek, A. M., *The Human Brain*. Springfield, Ill., Thomas, 1959.

Lenin, V. I., "The Proletarian Revolution and Kautsky," *Handbook of Marxism*. 1919.

————, *Selected Works*, Vols. 7 and 11. London, Lawrence & Wishart, 1936.

Lest We Forget! Washington, D.C., Government Printing Office, 1960.

Lewinsohn, Richard, *A History of Sexual Customs.* New York, Harper, 1958.

Linton, Ralph, *Science of Man in the World Crisis.* New York, Columbia, 1945.

Lippmann, Walter, *The Good Society.* New York, Grosset and Dunlap, 1943.

Luck, J. M., "Man Against His Environment," *Science.* Vol. 126, November 1, 1957.

MacCurdy, G. G., *The Coming of Man.* New York, University Society, 1935.

Macpherson, R. K., *Science News Letter,* Vol. 74. November, 1958.

Malthus, Thomas Robert, *An Essay on the Principle of Population.* New York, Dutton, 1933.

————, *The Principle of Population,* Vol. I, London, Dent, 1914.

Manly, Chesley, *One Billion Years.* Chicago, Natural History Museum.

Marshall, T. H., *et. al., The Population Problem.* London, Allen, 1938.

Marx, Karl, *The Correspondence of Marx and Engels.* New York, International, 1935.

Mathiez, Albert, *The French Revolution.* New York, Russell, 1956.

Mayr, Ernst, *Animal Species and Evolution.* Cambridge, Harvard, 1963.

————, "Comments," *Daedalus.* January 7, 1955.

McDougall, William, *Group Mind.* London, Cambridge, 1939.

McKelvey, V. B., *Science.* April 3, 1959.

Mead, Margaret, *Male and Female.* New York, New American Library, 1949.

Medical World News. February 28, 1964.

Melgounov, Sergey Petrovich, *The Red Terror in Russia.* London, Dent, 1926.

Michelmore, Susan, *Sexual Reproduction.* New York, Natural History, 1964.

Miller, Arjay, *Fortune.* August, 1968.

Miller, W. G., Robert, H. L., and Shulman, M. D., *The Meaning of Communism.* Morristown, N. J., Silver Burdette, 1963.

Montagu, M. F. Ashley, *Anthropology and Human Nature.* Boston, Sargent, 1957.

Morton, Frederick, *The Rothschilds.* Greenwich, Fawcett, 1961.

Muller, Hermann J., *The Future Physical Development of Man.* Penn. State, April, 1960.

————, "How Radiation Changes the Genetic Constitution," *Bulletin of Atomic Scientists.* 1955.

————, *Human Genetic Betterment.* American Humanist Association, 1963.

————, "Life," *Science.* January 7, 1955.

————, *The Next Hundred Years.* Seagram Symposium, November 22, 1957.

————, "Our Load of Mutations," *American Journal of Human Genetics.* Vol. 2, No. 2, June, 1950.

————, *Out of the Night.* New York, Vanguard, 1935.

—————, "The Prospects of Genetic Change," *American Scientist*. Vol. 47, December, 1959.
Munn, Norman L., "The Evolution of Mind," *Scientific American*. June, 1957.

Nations Business. Vol. 54, January, 1966.
Nel, Elizabeth, *Mr. Churchill's Secretary*. New York, Coward-McCann, 1958.
Nevins, Allen and Hill, Frank, "Ford:—," *Life*. July 13, 1959.
Noble, John, *I Was a Slave in Russia*. New York, Devin-Adair, 1958.

Oparin, A. I., *The Origin of Life*. New York, Dover Publications, 1938.
O'Riley, John, *Wall Street Journal*. Vol. 78, No. 35, February 19, 1958.
Osborn, Frederick, *The Future of Human Heredity*. New York, Weybright and Talley, 1968.
—————, "Galton and Mid-Century Eugenics," *Eugenics Review*. April, 1956.
—————, *Preface to Eugenics*. New York, Harper, 1951.
—————, *Science*. Vol. 125, March 22, 1957.
Osborn, Henry Fairfield, *Men of the Old Stone Age*. New York, Scribner's, 1918.

Pasvolsky, Leo, "The Intelligentsia Under the Soviets," *Atlantic Monthly*. Vol. 126, November, 1920.
Pauli, W. F., *World of Life*. New York, Houghton Mifflin, 1949.
Pauling, Linus, "Molecular Disease," *American Journal of Orthopsychiatry*. Vol. XXIX, 1959.
—————, "Molecular Disease and Evolution," *Proceedings of the Rudolf Virchow Medical Society in the City of New York*, Vol. 21.
Pauling, Linus and Itano, Harvey A., eds., *Molecular Structure and Biological Specificity*. Washington, D.C., American Institute of Biological Sciences, 1957.
Pendell, Elmer, *Sex vs. Civilization*. Los Angeles, Noontide, 1967.
Physics Today. May, 1965.
Plato, *The Republic*. Cleveland, World, 1946.
Platt, Jno. R., "The Step to Man," *Science*. Vol. 149, August 6, 1965.
Popenoe, Paul and Johnson, Roswell Hill, *Applied Eugenics*. New York, Macmillan, 1933.
Pyke, Margaret, "Family Planning," *Eugenics Review*. Vol. 55, July, 1963.

Reed, Charles A., "Animal Domestication," *Science*. Vol. 130, December, 1959.
Reed, Sheldon C., "The Evolution of Human Intelligence," *American Scientist*, Vol. 53, 1965.
Research Institute Report. July 2, 1965.
Rickover, H. G., "The Talented Mind," *Science News Letter*. March 16, 1957.
Riddle, Oscar, *The Unleashing of Evolutionary Thought*. New York, Vantage, 1954.
Russell, Lord Bertrand, *Marriage and Morals*. New York, Bantam, 1929.

Saleeby, Caleb Williams, *Parenthood and Race Culture*. New York, Moffat, 1909.

Santayana, George, *Life of Reason*, Vol. I. New York, Scribner's, 1932.

Saturday Evening Post, Vol. 223, August 6, 1960.

————, Vol. 236, January 26, 1963.

————, Vol. 239, December 3, 1966.

Scheinfeld, Amram, *The Basic Facts of Human Heredity*. New York, Washington Square, 1961.

Schwartz, Benjamin, "The Intelligentsia in Communist China," *Daedalus*. Summer, 1960.

Science News Letter. Vol. 73, June 1, 1957.

Scientific American. Vol. 200, January, 1959.

————. Vol. 207, September, 1962.

Scudder, Thayer, *The Next Ninety Years*. Pasadena, C. I. T., 1967.

Serebriakoff, Victor. *I.Q.—A Mensa Analysis*. Tiptree, Essex, Anchor, 1965.

Shaw, George Bernard, *Sociological Papers*. London, Macmillan, 1904.

Sherman, J. K., "Improved Methods of Preservation of Human Spermatozoa," *Fertility and Sterility*. Vol. 14, No. 1, January-February, 1963.

Sjoberg, Gideon, "The Origin and Evolution of Cities," *Scientific American*. Vol. 213, September, 1965.

Snyder, Laurance H., *The Principles of Heredity*. Boston, D. C. Heath, 1935.

Snyder, Louis L., *The War*. New York, Messner, 1960.

————, *The World in the 20th Century*. Princeton, Van Nostrand, 1964.

Sollas, William J., *Ancient Hunters*, 3rd ed. New York, Macmillan, 1924.

Souvarine, Boris, *Stalin*. Toronto, Longmans, 1939.

Soviet World Outlook. U.S. Dept. of State Publication 6836, 1959.

Speiser, E. A., "Ancient Mesopotamia," *National Geographic*. January, 1951.

Spengler, Oswald, *Decline of the West*, Vol. II. New York, Knopf, 1928.

Stalin, Joseph, *Leninism*. New York, International, 1933.

Stevers, Martin, *Mind Through the Ages*. Garden City, Doubleday, 1940.

Stoddard, Theodore, *The Revolt Against Civilization*. New York, Scribner's, 1922.

————, *The Rising Tide—*. New York, Scribner's, 1922.

Strohm, John, "How They Hate Us in Red China," *Reader's Digest*. Vol. 74, January, 1959.

Sumner, William Graham, *Essays*, Vol. II. New Haven, Yale, 1934.

Terry, Benjamin S., *The American Journal of Sociology*. Vol. II, 1896.

Thiers, Louis Adolphe, *The History of the French Revolution*, Vol. III. Philadelphia, Lippincott, 1894.

Thomson, Sir Godfrey, *The Trend of National Intelligence*. London, Eugenics Society, 1947.

————, *The Trend of Scottish Intelligence*. London, University of London Press, 1949.

Tilney, Frederick, *The Brain From Ape to Man*. Vol. II. New York, Hoeber, 1928.

————, *The Master of Destiny.* Garden City, Doubleday, 1930.
Tobias, Phillip V., "Physique of a Desert Folk," *Natural History.* Vol. 70. February, 1961.
Tocqueville, Alexis de, *The European Revolution.* Garden City, Doubleday, 1959.
Toynbee, Arnold J., *A Study of History.* New York and London, Oxford, 1946.
Toynbee, Arnold, *Lectures on the Industrial Revolution in England.* Boston, Beacon, 1956.
Tse-tung, Mao, "Problems of War and Strategy," *Chinese Communist World Outlook.* U.S. Dept. of State Publication 7379, 1962.
Tyler, Leona E., *The Psychology of Human Differences,* 2nd ed. New York, Appleton, 1956.

U. N. Demographic Yearbook. New York, 1963, 1965.
Upton, Miller, *U.S. News and World Report.* July 10, 1967.
U.S. News. Vol. 44, January 31, 1958.
————. Vol. 45, August 29, 1958.
————. Vol. 51, July 24, 1961.
————. Vol. 53, August 20, 1962.
————. Vol. 57, December 19, 1964.
————. Vol. 58, February 8, 1965.
————. Vol. 58, March 22, 1965.
————. Vol. 59, July 12, 1965.
————. Vol. 60, April 11, 1966.
————. Vol. 66, February 17, 1969.
Useller, James W., *Ordnance.* Vol. 43, September-October, 1958.
Urey, H. C., "On the Early Chemical History of the Earth and the Origin of Life," *Proc. Nat. Acad. Sci.,* 38, Washington, D.C., 1952.

Wall Street Journal, Pacific Coast Edition. Vol. LXXXIX, September 16, 1968.
Warden, Carl J., *Evolution of Human Behavior.* New York, Macmillan, 1932.
Weidenreich, Franz, "The Duration of Life in Fossil Man," *Chinese Medical Journal.* Washington, D.C., 1939.
Weir, John, *The Next Ninety Years.* Pasadena, Calif. Institute of Technology, 1967.
Wendt, Herbert, *In Search of Adam.* Boston, Houghton Mifflin, 1955.
Weyl, Nathaniel and Possony, Stefan T., *The Geography of Intellect.* Chicago, Regnery, 1963.
Whetham, C. D., "Decadance and Civilization," *Hibbert Journal.* Vol. 10, October, 1911.
White, Robert W., *Scientific American.* Vol. 201, September, 1959.
Williams, J. D., *Saturday Evening Post.* Vol. 223, August 6, 1960.
Willkie, Wendell L., *One World.* New York, Simon & Schuster, 1943.
Wolstenholme, Gordon, *Man and His Future.* Boston, Little Brown, 1963.
Woolley, Sir Charles Leonard, *The Sumerians.* Clarendon, Oxford, 1929.

Young, Clarence W. and Stebbins, G. Ledyard, *The Human Organism and the World of Life.* New York, Harpers, 1938.
Yutang, Lin, "Confucius and Marx," *The World's Great Religions,* Vol. I. New York, Time-Life, 1957.

Foundation for Advancement of Man

F.A.M. is a charitable, educational, non-profit foundation, established to carry out, and to help others carry out, some of the objectives set forth in this book.

Within the limits of its resources,* F.A.M. stands ready to encourage the increase of exceptionally capable minds by all proper means, including those outlined as Opportunities of the Intelligent. It also stands ready to counsel and cooperate with others who wish to organize for similar purposes.

*All royalties from the sale of this book go to F.A.M.

To the Foundation for Advancement of Man

Post Office Box 2480-D
Pasadena, California 91105, U.S.A.

May I have fuller information regarding the
Foundation for Advancement of Man? ....
I should like to assist with funds. ....
I should like to assist with services. ....
I should like to assist with counsel. ....
My marriage partner and I have proven our
ability to have very intelligent children.
We would, if given economic assistance,
be willing to have more. ....